PUBLISHER COMMENTARY

GAO Schedule Assessment Guide (GAO-16-89G) December 2015

This schedule guide is a companion to the Cost Estimating and Assessment Guide (GAO-09-3SP). A cost estimate cannot be considered credible if it does not account for the cost effects of schedule slippage. An effective methodology for developing, managing, and evaluating capital program cost estimates includes the concept of scheduling the necessary work to a timeline, as discussed in the Cost Guide. Typically, schedule variances are followed by cost variances and management tends to respond to schedule delays by adding more resources or authorizing overtime. Therefore, a reliable schedule can contribute to an understanding of the cost impact if the program does not finish on time. Further, a schedule risk analysis allows for program management to account for the cost effects of schedule slippage when developing the life-cycle cost estimate.

Having managed many construction projects over the years in the U.S., Europe and the Middle East, I can tell you that without a good (reasonable) schedule, a project cannot stay within budget. Budget is more important than schedule as far as I am concerned but without a well thought out schedule, the budget will be busted. Schedule slippage is bound to happen on any project due to unforeseen circumstances such as weather and politics, but I have a problem with circumstances that should have been foreseen. As a manager of project managers, I always took the approach that if something goes wrong, it is the Project Manager's fault. Many PMs working for me over the years felt this was unfair, but that's how it is. If there is a delay, the PM should have anticipated that and figured out a work-around. I don't give bonuses to projects that are behind schedule. I recommend you do the same. They will get over it.

Why buy a book you can download for free? We print this book so you don't have to.

First you gotta find a good clean (legible) copy and make sure it's the latest version (not always easy). Some documents found on the web are missing some pages or the image quality is so poor, they are difficult to read. We look over each document carefully and replace poor quality images by going back to the original source document. We proof each document to make sure it's all there – including all changes. If you find a good copy, you could print it using a network printer you share with 100 other people (typically its either out of paper or toner). If it's just a 10-page document, no problem, but if it's 250-pages, you will need to punch 3 holes in all those pages and put it in a 3-ring binder. Takes at least an hour.

It's much more cost-effective to just order the latest version from www.Amazon.com

This material is published by 4th Watch Publishing Co. We publish tightly-bound, full-size books at 8 ½ by 11 inches, with large text and glossy covers. 4th Watch Publishing Co. is a Service Disabled Veteran Owned Small Business (SDVOSB). Please visit www.usgovpub.com.

Other books available on www.Amazon.com :

GAO Green Book - Standards for Internal Control in the Federal Government

GAO Yellow Book - Government Auditing Standards

GAO Financial Audit Manual

DoD 7000.14 - R Financial Management Regulation

Defense Acquisition Guidebook (Chapters 1 - 10)

Federal Acquisition Regulation - Complete

Defense Federal Acquisition Regulation – Complete

OMB No. A-123 - Management's Responsibility for Enterprise Risk Management and Internal Control

OMB A-130 & Federal Information Security Modernization Act (FISMA)

Federal Information System Controls Audit Manual (FISCAM)

GAO Technology Readiness Assessment Guide

GAO Cost Estimating and Assessment Guide

SCHEDULE ASSESSMENT GUIDE

Best Practices for Project Schedules

GAO-16-89G

December 2015

PREFACE

The U.S. Government Accountability Office is responsible for, among other things, assisting the Congress in its oversight of the federal government, including agencies' stewardship of public funds. To use public funds effectively, the government must employ effective management practices and processes, including the measurement of government program performance.

Toward these objectives, in March 2009, we published the GAO *Cost Estimating and Assessment Guide* as a consistent methodology based on best practices that can be used across the federal government to develop, manage, and evaluate capital program cost estimates. The methodology outlined in the *Cost Estimating and Assessment Guide* is a compilation of best practices that federal cost estimating organizations and industry use to develop and maintain reliable cost estimates throughout the life of an acquisition program.

Thus, a well-planned schedule is a fundamental management tool that can help government programs use public funds effectively by specifying when work will be performed in the future and measuring program performance against an approved plan. Moreover, as a model of time, an integrated and reliable schedule can show when major events are expected as well as the completion dates for all activities leading up to them, which can help determine if the program's parameters are realistic and achievable.

Additionally, a well-formulated schedule can facilitate an analysis of how change affects the program. Accordingly, a schedule can serve as a warning that a program may need an overtarget budget or schedule.

The GAO *Schedule Assessment Guide* develops the scheduling concepts introduced in the *Cost Estimating and Assessment Guide* and presents them as ten best practices associated with developing and maintaining a reliable, high-quality schedule. The GAO *Schedule Assessment Guide* also presents guiding principles for auditors to evaluate certain aspects of government programs.

We intend to update the *Schedule Assessment Guide* to keep it current. Comments and suggestions from experienced users, as well as recommendations from experts in the scheduling, cost estimating, and program acquisition disciplines, are always welcome. If you have any questions concerning this guide, you may contact me at (202) 512-6412 or personst@gao.gov. Contact points for our Office of Congressional Relations and Office of Public Affairs may be found on the last page of this guide. Major contributors to this project are listed in appendix X.

T. M. Persons

Timothy M. Persons, Ph.D.
Director, Center for Science, Technology, and Engineering
Applied Research and Methods

CONTENTS

Preface i

Acronyms and Abbreviations xi

Introduction 1

Concepts 3

Ten Best Practices 3

The Integrated Master Schedule 5

The Critical Path Method 6

Planning, Scheduling, and the Scheduler 7

A Process for Creating and Maintaining Reliable Schedules 8

Best Practice 1: Capturing All Activities 11

Capturing All Effort 11

Constructing an IMS 15

The IMS as a Consolidation Tool 18

Work Breakdown Structure 20

Activity Names 22

Activity Codes 24

Best Practices Checklist: Capturing All Activities 25

Best Practice 2: Sequencing All Activities 27

Predecessor and Successor Logic 28

Early and Late Dates 30

Incomplete and Dangling Logic 31

Summary Logic 34

Date Constraints 36

Using Date Constraints 38

Lags and Leads 41

Using Lags...42

Using Leads..44

Path Convergence...46

Best Practices Checklist: Sequencing All Activities..46

Best Practice 3: Assigning Resources to All Activities.......................................**49**

Resources, Effort, and Duration..49

Rolling Wave Planning..51

Loading Activities with Resources..52

Resource Leveling..57

Best Practices Checklist: Assigning Resources to All Activities................................61

Best Practice 4: Establishing the Duration of All Activities................................**63**

Estimating Durations...65

Calendars..66

Best Practices Checklist: Establishing the Duration of All Activities.......................68

**Best Practice 5: Verifying That the Schedule Can Be Traced Horizontally
and Vertically**...**71**

Vertical Traceability...72

Best Practices Checklist: Verifying That the Schedule Can Be Traced Horizontally
and Vertically...73

Best Practice 6: Confirming That the Critical Path Is Valid................................**75**

The Critical Path and the Longest Path..77

Common Barriers to a Valid Critical Path..79

Resource Leveling and Critical Resources...86

Critical Path Management..87

Program and Project Critical Paths...88

Best Practices Checklist: Confirming That the Critical Path Is Valid.......................89

Best Practice 7: Ensuring Reasonable Total Float...**91**

Definitions of Total Float and Free Float..92

Calculating Float...93

Common Barriers to Valid Float..93

Reasonableness of Float...94

Float Management 95

Best Practices Checklist: Ensuring Reasonable Total Float 96

Best Practice 8: Conducting a Schedule Risk Analysis 99

Definition of Schedule Risk Analysis...99

Schedule Uncertainty and Risk..100

Merge Bias and Schedule Underestimation..100

Conducting a Schedule Risk Analysis 102

Collecting Anonymous and Unbiased Risk Data.................................105

Schedule Risk Analysis with Three-Point Duration Estimates..............106

Schedule Risk Analysis with Risk Drivers 108

Prioritizing Risks 110

Probabilistic Branching 113

Correlation...115

Schedule Contingency...117

Updating and Documenting a Schedule Risk Analysis.........................118

Best Practices Checklist: Conducting a Schedule Risk Analysis.............119

Best Practice 9: Updating the Schedule Using Actual Progress and Logic 121

Statusing Progress...122

Progress Records...125

Adding and Deleting Activities...126

Out-of-Sequence Logic 126

Verifying Status Updates and Schedule Integrity.................................129

Schedule Narrative 130

Reporting and Communication 132

Best Practices Checklist: Updating the Schedule Using Actual Progress and Logic.....132

Best Practice 10: Maintaining a Baseline Schedule 135

Baseline and Current Schedules..135

Basis Document 137

The Change Process...138

Baseline Analysis 140

Trend Analysis..143

Strategies for Recovery and Acceleration...145

Best Practices Checklist: Maintaining a Baseline Schedule.................146

Four Characteristics of a Reliable Schedule **148**

Appendix I: Objectives, Scope, and Methodology..............................**150**

Appendix II: An Auditor's Key Questions and Documents **151**

Appendix III: Scheduling and Earned Value Management **166**

Appendix IV: The Forward and Backward Pass **172**

Appendix V: Common Names for Schedule Date Constraints and Their Effects...**181**

Appendix VI: Standard Quantitative Measurements for Assessing Schedule Health...**183**

Appendix VII: Comparison of GAO's Schedule Assessment Guide to Key Industry and Agency Schedule Guidance.......................................**189**

Appendix VIII: Recommended Elements of a Data Collection Instrument..........**197**

Appendix IX: Case Study Backgrounds...**198**

Appendix X: Experts Who Helped Develop This Guide **207**

Appendix XI: GAO Contacts and Staff Acknowledgments..................**215**

Glossary...**216**

References **220**

TABLES

Table 1. A Process for Creating and Maintaining Reliable Schedules 9

Table 2: Estimated Durations for a Section of the House Schedule 106

Table 3: Some Identified Risks for a House Construction Schedule 109

Table 4: Some Uncertainties for a House Construction Schedule 109

Table 5: Top Prioritized Risks in the House Construction Schedule 113

Table 6: Strategies for Recovery and Acceleration 145

Table 7: Best Practices Entailed in the Four Characteristics of a Reliable Schedule..... 149

Table 8: Seven Measures of Effort 170

Table 9: Expected Durations and Estimated Resources in House Construction 173

Table 10: Common Names for Date Constraints and Their Primary Effects 181

Table 11: Standard Data Measures for Schedule Best Practices 183

Table 12. Generally Accepted Scheduling Principles and GAO Best Practices
Compared 195

Table 13: Case Studies Drawn from GAO Reports Illustrating This Guide 198

FIGURES

Figure 1. A Process for Creating and Maintaining Reliable Schedules 8

Figure 2: A Milestone's Dates 13

Figure 3: Detail Activities 14

Figure 4: Summary Activities 15

Figure 5: The WBS and Scheduling 20

Figure 6: Redundant Activity Names 23

Figure 7: Unique Activity Names 23

Figure 8: A Finish-to-Start Relationship 28

Figure 9: A Start-to-Start Relationship 29

Figure 10: A Finish-to-Finish Relationship 29

Figure 11: Early and Late Dates 31

Figure 12: Start-Date Dangling Logic 32

Figure 13: Start-Date Dangling Logic Corrected with Predecessor Logic 33

Figure 14: Finish-Date Dangling Logic 33

Figure 15: Finish-Date Dangling Logic Corrected with Successor Logic 34

Figure 16: A Linked Summary Activity 35

Figure 17: A Lag...42

Figure 18: A Negative Lag (or Lead) 42

Figure 19: Using a Lag to Force an Activity's Start Date 42

Figure 20: The Effect of a Lag on Successor Activities......................................43

Figure 21: Eliminating Leads with Finish-to-Start Links..................................44

Figure 22: Logic Failure Associated with a Lead 44

Figure 23: Enumerating Lags 45

Figure 24: A Profile of Expected Construction Labor Costs by Month 55

Figure 25: A Smoothed Profile of Expected Construction Labor Costs by Month 56

Figure 26: Resource Overallocation in a Correctly Sequenced Network............59

Figure 27: Resource Leveling in a Correctly Sequenced Network......................60

Figure 28: The Critical Path and Total Float 76

Figure 29: The Critical Path and the Longest Path 78

Figure 30: Critical Path Activities Not on the Longest Path.............................78

Figure 31: The Longest Path and the Lowest-Float Path 79

Figure 32: The Effect of Multiple Calendars on the Critical Path.....................82

Figure 33: The Critical Path and Lags...84

Figure 34: An Incorrect Critical Path with Level-of-Effort Activities...............85

Figure 35: Total Float and Free Float 93

Figure 36: A Simple Schedule as a Network Diagram 101

Figure 37: The Cumulative Distribution of the House Construction Schedule..........107

Figure 38: House Construction Schedule Results from a Risk Driver Simulation 110

Figure 39: Sensitivity Indexes for the House Construction Schedule 111

Figure 40: Evaluation of Risk Sensitivity in the House Construction Schedule 111

Figure 41: Risk Criticality of Selected Activities in the House Construction
Schedule 112

Figure 42: Probabilistic Branching in a Schedule..114

Figure 43: Probability Distribution Results for Probabilistic Branching....................115

Figure 44: Probability Distribution Results for Risk Analysis with and without
Correlation 116

Figure 45: The Original Plan for Interior Finishing 127

Figure 46: Retained Logic...128

Figure 47: Progress Override 128

Figure 48: Baselined Activities 141

Figure 49: Updated Status Compared with Baseline 141

Figure 50: Updated Status and Proposed Plan Compared with Baseline 143

Figure 51: Start-Up and Testing Network 173

Figure 52: Early Start and Early Finish Calculations 174

Figure 53: Successive Early Start and Early Finish Calculations 175

Figure 54: Complete Early Start and Early Finish Calculations 176

Figure 55: Late Start and Late Finish Calculations 177

Figure 56: Early and Late Dates of a Start-Up and Testing Network 178

Figure 57: Total Float in a Start-Up and Testing Network 179

Contents

CASE STUDIES

Case Study 1: Attempts in Varying Degrees to Capture All Effort, from *DOD Business Transformation,* GAO-11-53 12

Case Study 2: Missing Increments, from *DOD Business Systems Modernization,* GAO-14-152.................18

Case Study 3: Consolidated program schedules, from *Arizona Border Surveillance Technology Plan,* GAO-14-368 19

Case Study 4: Successfully Coding Detailed Work, from *VA Construction,* GAO-10-189 25

Case Study 5: Summary Logic, from *DOD Business Transformation,* GAO-11-53.................................35

Case Study 6: Managing Resources with Constraints, from *DOD Business Transformation,* GAO-11-53.....................39

Case Study 7: Assigning Resources to Activities, from *Nuclear Nonproliferation,* GAO-10-378 54

Case Study 8: Assuming Unlimited Availability of Resources, from *Arizona Border Surveillance Technology Plan,* GAO-14-368 57

Case Study 9: The Effect of Incorrect Calendars, from *Transportation Worker Identification Credential,* GAO-10-43 68

Case Study 10: Missing Vertical Traceability, from *DOD Business Systems Modernization,* GAO-14-152.....................73

Case Study 11: Noncontinuous Critical Path, from *FAA Acquisitions,* GAO-12-223 81

Case Study 12: Predetermined Critical Activities, from *Immigration Benefits,* GAO-12-66 86

Case Study 13: Unreasonable Float from the Sequencing of Activities, from *FAA Acquisitions,* GAO-12-223.............94

Case Study 14: Converging Paths and Schedule Risk Analysis, from *Coast Guard,* GAO-11-743 102

Case Study 15: Schedule Risk Analysis, from *VA Construction,* GAO-10-189..104

Case Study 16: Invalid Status Dates, from *DOD Business Transformation,* GAO-11-53 123

Case Study 17: Updating a Schedule Using Progress and Logic, from *Aviation Security,* GAO-11-740 125

Case Study 18: Inconsistent Documentation Requirements, from *Polar-orbiting Environmental Satellites,* GAO-13-676..131

Case Study 19: Baseline Schedule, from *2010 Census,* GAO-10-59 139

ACRONYMS AND ABBREVIATIONS

BEI	baseline execution index
BOE	basis of estimate
CLIN	contract line item number
CPM	critical path method
DEAMS	Defense Enterprise Accounting and Management System
DHS	U.S. Department of Homeland Security
DOE	U.S. Department of Energy
ERP	enterprise resource planning
EVM	earned value management
FAA	Federal Aviation Administration
FNET	finish no earlier than
FNLT	finish no later than
FTE	full-time equivalent
GCSS-Army	Army Global Combat Support System
GFEBS	General Fund Enterprise Business System
G/R	giver/receiver
IMS	integrated master schedule
IPT	integrated product team
LCCE	life-cycle cost estimate
LOE	level of effort
MFON	must finish on
MSON	must start on
PMB	performance measurement baseline
PMO	program management office
PWS	performance work statement
QBD	quantifiable backup data
SNET	start no earlier than
SNLT	start no later than
SOO	statement of objectives
SOW	statement of work
TSA	Transportation Security Administration
TPO	Transformation Program Office
USCIS	U.S. Citizenship and Immigration Services
WBS	work breakdown structure

INTRODUCTION

The success of a program depends in part on having an integrated and reliable master schedule that defines when and how long work will occur and how each activity is related to the others. A schedule is necessary for government acquisition programs for many reasons. The program schedule provides not only a road map for systematic project execution but also the means by which to gauge progress, identify and resolve potential problems, and promote accountability at all levels of the program. A schedule provides a time sequence for the duration of a program's activities and helps everyone understand both the dates for major milestones and the activities that drive the schedule. A program schedule is also a vehicle for developing a time-phased budget baseline.

Moreover, the schedule is an essential basis for managing tradeoffs between cost, schedule, and scope. Among other things, scheduling allows program management to decide between possible sequences of activities, determine the flexibility of the schedule according to available resources, predict the consequences of managerial action or inaction in events, and allocate contingency plans to mitigate risk. Following changes in a program, the schedule is used to forecast the effects of delayed, deleted, and added effort, as well as possible avenues for time and cost recovery. In this respect, schedules can be used to verify and validate proposed adjustments to the planned time to complete.

The GAO *Schedule Assessment Guide* is intended to expand on the scheduling concepts introduced in the *Cost Estimating and Assessment Guide* by providing ten best practices to help managers and auditors ensure that the program schedule is reliable. The reliability of the schedule determines the credibility of the program's forecasted dates for decision making.

Our approach to developing this guide was to ascertain best practices from leading practitioners and to develop standard criteria to determine the extent agency programs and projects meet industry scheduling standards. To develop criteria for scheduling standards, we expanded on the criteria originally published in GAO's *Cost Estimating and Assessment Guide*. We developed each best practice in consultation with a committee of cost estimating, scheduling, and earned value analysis specialists from across government, private industry, and academia. We released a public exposure draft of the GAO *Schedule Assessment Guide* in May 2012 and sought input and feedback from all who expressed interest for two years. We also compared the standards detailed in the guide with schedule standards and best practices developed by other agencies and organizations. We describe our scope and methodology in detail in appendix I. Some case studies in this guide are reprinted from GAO reports that are several years old. These case studies are reflective of agency practices at the time and are provided for illustration purpose only.

We conducted our work from November 2010 to November 2015 in accordance with all sections of GAO's Quality Assurance Framework that are relevant to our objectives. The framework requires that we plan and perform the engagement to obtain sufficient and appropriate evidence to meet our stated objectives and to discuss any limitations in our work. We believe that the information and data obtained, and the analysis conducted, provide a reasonable basis for the guidance in this product.

CONCEPTS

TEN BEST PRACTICES

The ten best practices associated with a high-quality and reliable schedule and their concepts are as follows.

1. Capturing all activities. The schedule should reflect all activities as defined in the program's work breakdown structure (WBS), which defines in detail the work necessary to accomplish a project's objectives, including activities both the owner and the contractors are to perform.

2. Sequencing all activities. The schedule should be planned so that critical program dates can be met. To do this, activities must be logically sequenced and linked—that is, listed in the order in which they are to be carried out and joined with logic. In particular, a predecessor activity must start or finish before its successor. Date constraints and lags should be minimized and justified. This helps ensure that the interdependence of activities that collectively lead to the completion of activities or milestones can be established and used to guide work and measure progress.

3. Assigning resources to all activities. The schedule should reflect the resources (labor, materials, travel, facilities, equipment, and the like) needed to do the work, whether they will be available when needed, and any constraints on funding or time.

4. Establishing the duration of all activities. The schedule should realistically reflect how long each activity will take. When the duration of each activity is determined, the same rationale, historical data, and assumptions used for cost estimating should be used. Durations should be reasonably short and meaningful and should allow for discrete progress measurement. Schedules that contain planning and summary planning packages as activities will normally reflect longer durations until broken into work packages or specific activities.

5. Verifying that the schedule can be traced horizontally and vertically. The schedule should be horizontally traceable, meaning that it should link products and outcomes associated with other sequenced activities. Such links are commonly referred to as "hand-offs" and serve to verify that activities are arranged in the right order for achieving aggregated products or outcomes. The schedule should also be vertically traceable—that is, data are consistent between different levels of a schedule. When schedules are vertically traceable, lower-level schedules are clearly consistent with upper-level schedule milestones, allowing for total schedule integrity and enabling different teams to work to the same schedule expectations.

6. Confirming that the critical path is valid. The schedule should identify the program's critical path—the path of longest duration through the sequence of activities. Establishing a valid critical path is necessary for examining the effects of any activity's slipping along this path. The program's critical path determines the program's earliest completion date and focuses the team's energy and management's attention on the activities that will lead to the project's success.

7. Ensuring reasonable total float. The schedule should identify reasonable total float (or slack)—the amount of time a predecessor activity can slip before the delay affects the program's estimated finish date—so that the schedule's flexibility can be determined. The length of delay that can be accommodated without the finish date's slipping depends on the number of date constraints within the schedule and the degree of uncertainty in the duration estimates, among other factors, but the activity's total float provides a reasonable estimate of this value. As a general rule, activities along the critical path have the least total float. Unreasonably high total float on an activity or path indicates that schedule logic might be missing or invalid.

8. Conducting a schedule risk analysis. A schedule risk analysis starts with a good critical path method schedule. Data about program schedule risks are incorporated into a statistical simulation to predict the level of confidence in meeting a program's completion date; to determine the contingency, or reserve of time, needed for a level of confidence; and to identify high-priority risks. Programs should include the results of the schedule risk analysis in constructing an executable baseline schedule.

9. Updating the schedule using actual progress and logic. Progress updates and logic provide a realistic forecast of start and completion dates for program activities. Maintaining the integrity of the schedule logic is necessary to reflect the true status of the program. To ensure that the schedule is properly updated, people responsible for the updating should be trained in critical path method scheduling.

10. Maintaining a baseline schedule. A baseline schedule is the basis for managing the program scope, the time period for accomplishing it, and the required resources. The baseline schedule is designated the target schedule and is subjected to a configuration management control process. Program performance is measured, monitored, and reported against the baseline schedule. The schedule should be continually monitored so as to reveal when forecasted completion dates differ from baseline dates and whether schedule variances affect downstream work. A corresponding basis document explains the overall approach to the program, defines custom fields in the schedule file, details ground rules and assumptions used in developing the schedule, and justifies constraints, lags, long activity durations, and any other unique features of the schedule.

The ten best practices represent the key concepts of a reliable schedule. These best practices are in no particular order; they are not intended as a series of steps for developing the schedule.

The federal audit community is the primary audience for this guide. Agencies that do not have a formal policy for creating or maintaining integrated master schedules will also benefit from the guide because it will inform them of GAO's criteria for assessing a schedule's credibility. Besides GAO, auditing agencies include Inspectors General and agency audit services. Following the text discussion of the best practices, an appendix lists key questions and documentation that members of the federal audit community who assess program schedules will find useful. The remainder of this section introduces the concepts and activities entailed in the integrated master schedule; the critical path method; planning, scheduling, and the scheduler; and a process for creating and maintaining reliable schedules.

THE INTEGRATED MASTER SCHEDULE

As a document that integrates the planned work, the resources necessary to accomplish that work, and the associated budget, the IMS should be the focal point of program management. In this guide, an IMS constitutes a program schedule that includes the entire required scope of effort, including the effort necessary from all government, contractor, and other key parties for a program's successful execution from start to finish.[1]

An IMS connects all the scheduled work of the government and the contractor in a network, or collection of logically linked sequences of activities. The sequences clearly show how related portions of work depend on one another, including the relationships between the government and contractors. Although the IMS includes all government, contractor, and external effort, the government program management office is ultimately responsible for its development and maintenance. In this respect, the government program management office must ensure that the schedule is as logical and realistic as possible. The IMS must be a complete and dynamic network. That is, the IMS should consist of logically related activities whose forecasted dates are automatically recalculated when activities change. If the schedule is not dynamic, planned activities will not react logically to changes, and the schedule will not be able to identify the consequences of changes or possible managerial action to respond to them.

In general, schedules can refer to programs and projects. In this guide, a "program" encompasses an entire program from beginning to end, including all government and contractor effort. An IMS may be made up of several or several hundred individual schedules that represent portions of effort within a program. These individual schedules

---

[1]We recognize that different organizations may use the term "integrated master schedule" differently; for example, IMS is often used to refer solely to the prime contractor schedule. Our use of "integrated" implies the schedule's incorporation of all activities—those of the contractor and government—necessary to complete a program.

are "projects" within the larger program. For example, a program IMS may consist of individual project schedules for the prime contractor, the government program management office, and a government testing laboratory.

As discussed in Best Practice 1, the IMS includes summary, intermediate, and all detailed schedules. At the highest level, the summary schedule provides a strategic view of the activities and milestones necessary to start and complete a program. The intermediate schedule includes all information displayed in the summary schedule, as well as key program activities and milestones that show the important steps in achieving high-level milestones. At the lowest level, the detailed schedule lays out the logically sequenced day-to-day effort to reach program milestones. Ideally, one schedule serves as the summary, intermediate, and detailed schedule by simply rolling up lower levels of effort into summary activities or higher-level work breakdown structure (WBS) elements.

The program or project team should develop the schedules and, in doing so, include the program manager, schedulers, and subject matter experts or managers responsible for specific areas of work. Managers responsible for resources should approve the areas of a schedule they are committed to support. If the schedule is not planned in sufficient detail or collaboratively by team members and stakeholders, then opportunities for process improvement (for example, identifying redundant activities), what-if analysis, and risk mitigation will be missed. Moreover, activity owners responsible for managing the day-to-day effort and the most experienced team members who perform the work are the best source of resource estimates. Activity owners must be able to explain the logic behind their resource estimates; if resources are without justification, management will lack confidence in the estimated durations and the schedule may falsely convey accuracy.

THE CRITICAL PATH METHOD

The critical path method is used to derive the critical activities—that is, activities that cannot be delayed without delaying the end date of the program. The amount of time an activity can slip before the program's end date is affected is known as "total float."

Critical activities have the least amount of float and, therefore, any delay in them generally causes the same amount of delay in the program's end date. Activities with total float within a narrow range of the critical path total float are called "near-critical" activities, because they can quickly become critical if their small amount of total float is used up in a delay. Management must closely monitor critical and near-critical activities by using sound schedule practices.

Unless the IMS represents the entire scope of effort and the effort is correctly sequenced through network logic, the scheduling software will report an incorrect or invalid critical path. That is, the critical path will not represent the activities affecting the program finish date. With no accurate critical path, management cannot focus on the activities that will be detrimental to the program's key milestones and finish date if they slip.

PLANNING, SCHEDULING, AND THE SCHEDULER

Project planning is a process within program management. An integral stage of management, it results primarily in an overall program execution strategy. The overall strategy is documented in the project plan, which defines, among other things,

- project scope;
- project objectives and requirements;
- stakeholders;
- organizational and work breakdown structures;
- design, procurement, and implementation; and
- risk and opportunity management plans.

Project planning is the basis for controlling and managing project performance, including managing the relationship between cost and time.

Scheduling is a distinct process that follows the planning process. The schedule is essentially a model of the project plan. It calculates the dates on which activities are to be carried out according to the project plan. As a model of time, the schedule incorporates key variables such as nonworking calendar periods, contingency, resource constraints, and preferred sequences of work activities to determine the duration and the start and finish dates of activities and key deliverables.

Planning and scheduling are continual processes throughout the life of a project. Planning may be done in stages throughout the project as stakeholders learn more details. This approach to planning, known as rolling wave planning, is discussed in Best Practice 3. Scheduling involves the management and control of the schedule over the project's life cycle. However, in no case should planning be concurrent with scheduling. In other words, work and strategies for executing the work must be planned first before activities can be scheduled.

By creating and maintaining the schedule, a scheduler interprets and documents the project plan developed by those responsible for managing and for executing the work. The scheduler is responsible for creating, editing, reviewing, and updating the schedule and ensuring that project and activity managers follow a formal schedule maintenance process. Interpreting the project's sequence of work entails responsibility for alerting program management to threats to the critical path, the degradation of float, and the derivation and use of schedule contingency. These concepts are discussed in Best Practices 6, 7, and 8. The scheduler must also modify the schedule in accordance with rolling wave planning details and approved change requests, including changes in scope and resource constraints. Maintaining a reliable schedule allows the scheduler to identify the effects of delayed activities or unplanned events on the planned sequence of activities, as well as possible mitigation strategies to prevent significant delays in planned work. The scheduler must track and report actual work performance against the plan, including the production of variances, forecasts, and what-if analyses.

A Process for Creating and Maintaining Reliable Schedules

As we noted earlier, the best practices described in this guide are presented in no particular order. However, they can be mapped to an overall process of established methods that result in high-quality schedules. This process is presented in figure 1 and described in detail in table 1.

The process in figure 1 is cyclic and described by elaboration through the rolling wave process. As the program proceeds, more becomes known about the detail work that needs to be done; risks are discovered, mitigated, or realized; and effort may be added or reduced.

Figure 1. A Process for Creating and Maintaining Reliable Schedules

Start

Capture all actvities

Create logically sequenced activity network

Estimate work and durations and assign resources

Progressive elaboration

Update, revise, and manage change

Stakeholder involvement

Validate critical path and reasonable total float

Set and document baseline

Verify and validate traceability

Analyze schedule risk

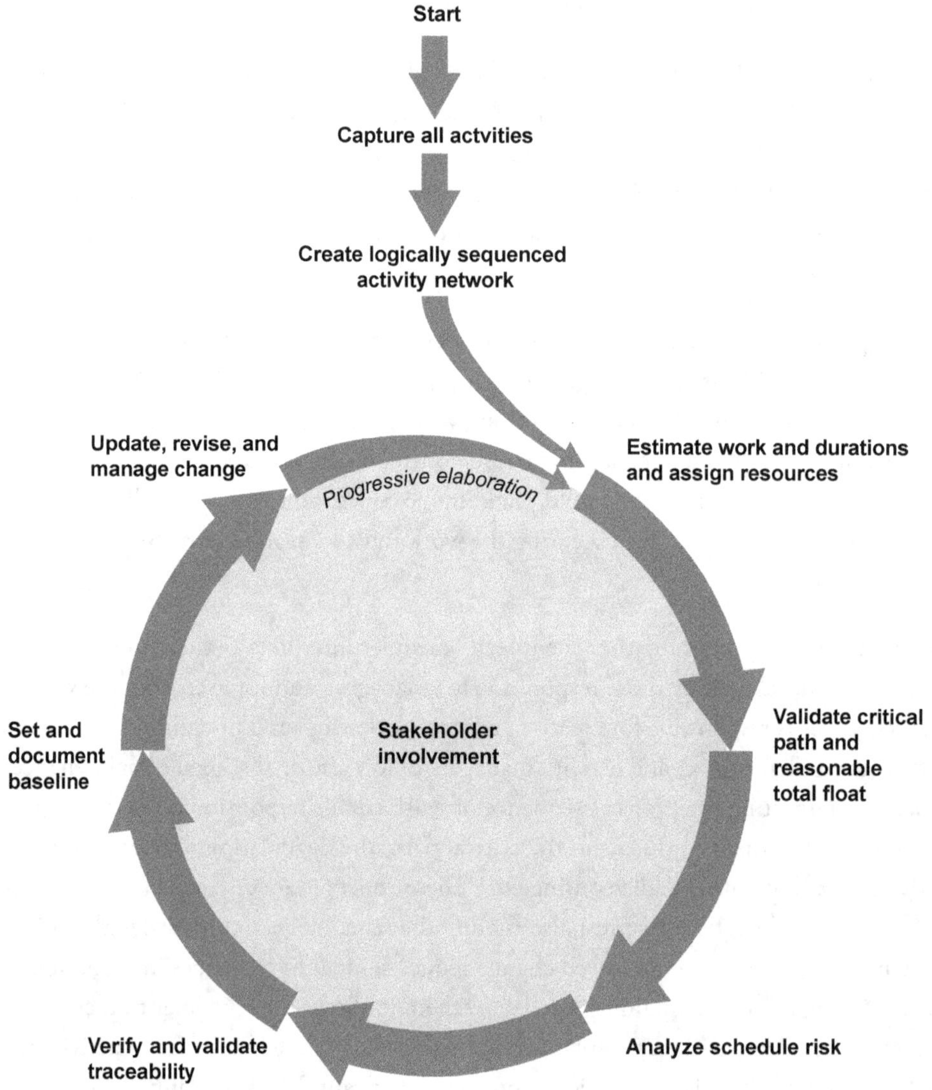

Source: Adapted from Keith D. Hornbacher. | GAO-16-89G

Table 1. A Process for Creating and Maintaining Reliable Schedules

Process step	Description	Corresponding scheduling best practice
Capture all activities	Using the work breakdown structure as a basis, all activities are captured in the schedule, including all work necessary by the owner and contractors. The schedule should reflect all effort necessary to accomplish the deliverables described in the WBS. Depending on how much is known, some sets of activities will be scheduled in detail and others will be planned in long-duration planning packages.	Capture all activities
Create logically sequenced activity network	Activities are listed in the order they are to be performed and are joined with logic to create predecessors and successors. Logic relationships are not made overly complex and date constraints and lags are minimized.	Sequence all activities
Estimate work and durations and assign resources	In accordance with rolling wave planning, estimates of work, duration, and effort are created for activities and resources are assigned. Budgets for direct labor, travel, equipment, material, and the like are assigned to both detail activities and planning packages so that total costs to complete the program are identified.	Assign resources to all activities; establish the duration of all activities
Validate critical path and reasonable total float	The critical and longest paths are identified and validated by the schedulers, management, and subject matter experts. Estimates of total float are examined for reasonableness and extreme values of float are confirmed after validating the network logic. Date constraints causing negative total float are examined and justified. The initial plan or updated schedule may need to be optimized. Strategies for recovery and acceleration can be used to allocate resources more efficiently and to meet time or cost constraints. Recovery options are created for significant forecasted delays.	Confirm that the critical path is valid; ensure reasonable total float
Analyze schedule risk	Data about program schedule risks are incorporated into a statistical simulation to predict the level of confidence in meeting a program's completion date; determine the contingency, or reserve of time, needed for a level of confidence; and identify high-priority risks and their mitigation plans. A schedule risk analysis is performed on the schedule before a baseline is set and periodically as the schedule is updated to reflect actual progress on activity durations and sequences.	Conduct a schedule risk analysis; capture all activities
Verify and validate traceability	The schedule's traceability, horizontally and vertically, is verified. Horizontal traceability ensures that products and outcomes are linked to associated activities. Vertical traceability ensures that data are consistent between different levels of the schedule.	Verify that the schedule can be traced horizontally and vertically
Set and document baseline	The baseline schedule is designated the target schedule and is subjected to configuration management control. A corresponding basis document explains the overall approach to the program and documents and justifies features of the schedule.	Maintain a baseline schedule

Process step	Description	Corresponding scheduling best practice
Update, revise, and manage change	Progress updates and logic provide a realistic forecast of start and completion dates for activities. The true status of the program is reflected through the integrity of the schedule logic. Performance is measured, monitored, and reported against the baseline schedule. The schedule is monitored to reveal when forecasted completion dates differ from baseline dates and whether schedule variances will affect downstream work. Trend analysis provides insight into program performance. Strategies for recovery and acceleration can be used to allocate resources more efficiently and to meet time or cost constraints. Recovery options are created for significant forecasted delays.	Update the schedule using actual progress and logic; maintain a baseline schedule

Source: GAO and Keith D. Hornbacher | GAO-16-89G.

The remainder of this document consists of detailed definitions and descriptions of the ten best practices.

BEST PRACTICE 1

Capturing All Activities

Best Practice 1: The schedule should reflect all activities as defined in the program's work breakdown structure (WBS), which defines in detail the work necessary to accomplish a program's objectives, including activities both the owner and contractors are to perform.

A schedule represents an agreement for executing a program. It should reflect all activities (for example, steps, events, required work, and outcomes) that will accomplish the deliverables described in the program's WBS. An IMS should be based on critical path method scheduling that contains all the work represented in logically linked activities representing the execution plan. At its summary level, the IMS gives a strategic view of activities and milestones necessary to start and complete a program. At its most detailed, the schedule clearly reflects the WBS and defines the activities necessary to produce and deliver each product. The detail should be sufficient to identify the longest path of activities through the entire program.

CAPTURING ALL EFFORT

The IMS should reflect all effort necessary to successfully complete the program, regardless of who performs it. Failing to include all work for all deliverables, regardless of whether they are the government's responsibility or the contractor's, can hamper program members' understanding the plan completely and the program's progressing toward a successful conclusion. If activities are missing from the schedule, then other best practices will not be met. Unless all necessary activities are accounted for, no one can be certain whether all activities are scheduled in the correct order, resources are properly allocated, the critical path is valid, or a schedule risk analysis will account for all risk.

Because the schedule is used for coordination, the absence of necessary elements will hinder coordination, increasing the likelihood of disruption and delay. A comprehensive IMS should reflect all a program's activities and recognize that uncertainties and unknown factors in schedule estimates can stem from, among other things, data limitations. A schedule incorporates levels of detail that depend on the information available at any point in time through a process known as rolling wave planning. Rolling wave planning is described in Best Practice 3.

A program IMS is not simply the prime contractor's schedule; it is a collection point for all work scopes executed by the program. That is, it is a comprehensive plan of all government, contractor, subcontractor, and key vendor work that must be performed. Along with complete contract life-cycle effort, the schedule must account for related government effort such as design reviews, milestone decisions, receipt of government-furnished equipment, and testing. It is also important to include the government effort that leads to the final acceptance of a product or service—for example, certain activities that only the government can perform, such as reviewing and accepting deliveries, obtaining permits, and performing program reviews.

Schedulers should be aware of how long these government activities take because they often have a clear effect on schedules; for instance, a program phase cannot begin until a government review is complete. In addition, if risk mitigation plans have been determined and agreed on, then mitigation activities should also be captured in the sched-

---

Case Study 1: Attempts in Varying Degrees to Capture All Effort, from *DOD Business Transformation*, GAO-11-53

GAO analyzed four enterprise resource planning system schedules and found that none of the programs had developed a fully integrated master schedule as an effective tool to help in the management of the programs. In particular, the schedules differed in the extent to which they captured all activities, as well as in their integration of government and contractor activities. For example, the Defense Enterprise Accounting and Management System Program Management Office did not have a schedule that integrated government and contractor activities. It maintained internal schedules that reflected government-only activities, but these activities were not linked to the contractor's activities. While the Army's Global Combat Support System schedule identified contractor activities, it contained only key government milestones for the program. Other government activities, such as testing events and milestones beyond December 2010, were not captured in the schedule. Instead, they were displayed in isolated, high-level illustrated documents. The Expeditionary Combat Support System program schedule contained detailed activities associated with government effort and contractor effort. However, the government activities were not fully linked to contractor activities, so that updates to government activities did not directly affect scheduled contractor activities. Finally, while the General Fund Enterprise Business System schedule captured government and contractor activities, key milestones in deployment, software release, and maintenance were not fully integrated, precluding a comprehensive view of the entire program.

In scheduling, best practices are interrelated so that deficiencies in one best practice cause deficiencies in other best practices. For example, if the schedule does not capture all activities, then there will be uncertainty about whether activities are sequenced in the correct order and whether the schedule properly reflects the resources needed to accomplish the work.

GAO, *DOD Business Transformation: Improved Management Oversight of Business System Modernization Efforts Needed*, GAO-11-53 (Washington, D.C.: October 7, 2010).

ule.[2] In particular, risk mitigation activities with scope and assigned resources should appear as discrete activities in the schedule.

A contractor project schedule, as a subset of the overall government program effort, includes only contractually authorized work because contractors are obligated to plan activities required by, and limited to, the contract. It is therefore the responsibility of the government program management office to integrate all government and contractor work—contractually authorized or not—into one comprehensive program plan that can be used to reliably forecast key program dates.

Moreover, everyone who is affected by the schedule should clearly agree on the final actions that constitute the completion of the program. For instance, if the scope includes financial closeout, contract disputes, and final payment activities, these should be completed before the finish milestone. Case study 1 gives examples of partially integrated schedules.

Milestone, Detail, and Summary Activities

Planned effort and events are represented in a schedule by a combination of milestones, detail activities, and summary activities. Milestones are points in time that have no duration but that denote the achievement or realization of key events and accomplishments such as program events or contract start dates. Because milestones lack duration, they do not consume resources. Two important milestones that every schedule should include are the project's start and its finish. No work should begin before the start milestone, and all project scope must be completed before the finish milestone. A project plan that does not emanate from a single start milestone activity and terminate at a single finish milestone activity is not properly constructed and may produce an erroneous critical path. Figure 2 is an example of a milestone.

Figure 2: A Milestone's Dates

Activity	Duration	Start	Finish	08/3/2025						
				S	M	T	W	T	F	S
Budget and design complete	0 days	8/8	8/8						◆	

Source: GAO. | GAO-16-89G

A best practice is to include milestones only to reflect major events or deliverables.[3] A milestone should have clear conditions for completion. Examples of milestones include the start and finish of the design stage, start and finish of subcontractor work, and key hand-off dates between parties. The presence of too many milestones in a schedule may

---

[2] More information on formal risk assessment is available in Best Practice 8, as well as in GAO, *Cost Estimating and Assessment Guide*, GAO-09-3SP (Washington, D.C.: Mar. 2009).

[3] For example, Federal Aviation Administration (FAA) service organizations employ standard program milestones when planning, executing, and reporting progress on investment programs. The standard program milestones are documented along with a description, completion criteria, WBS reference or crosswalk, and the decision authority.

mask the activities necessary to achieve key milestones and may prevent the proper recording of actual progress. That is, when too many milestones are introduced into a schedule, the activity sequences that are most likely to delay milestone achievement may become increasingly difficult to identify. If work is represented by milestones, actual progress recorded in the schedule cannot be used to forecast the dates of key events.

Detail activities are at the lowest level of the WBS and represent the performance of actual discrete work that is planned in the program. They are measurable portions of effort that result in a discrete product or component. They are also logically linked to other preceding and succeeding activities to form logical sequences and parallel paths of work that must be accomplished to complete the program. Logically related paths of detail activities are linked to milestones to show the progression of work that is planned. Detail activities have an estimated duration—that is, a planned estimate of the time it will take to complete the work—and are assigned resources. The status of detail activities is examined regularly to record actual progress. Figure 3 shows an example of a sequence of activities necessary to complete framing in a house construction project.[4]

Figure 3: Detail Activities

Activity	Duration	Start	Finish
Set steel columns and beams	1 day	10/22	10/22
Install first floor joists and decking	2 days	10/23	10/24
Frame first floor walls; install wall sheathing	6 days	10/27	11/3
Install roof trusses	2 days	11/4	11/5
Install roof decking	2 days	11/6	11/7
Inspect rough-in framing	1 day	11/10	11/10
Framing complete	0 days	11/10	11/10

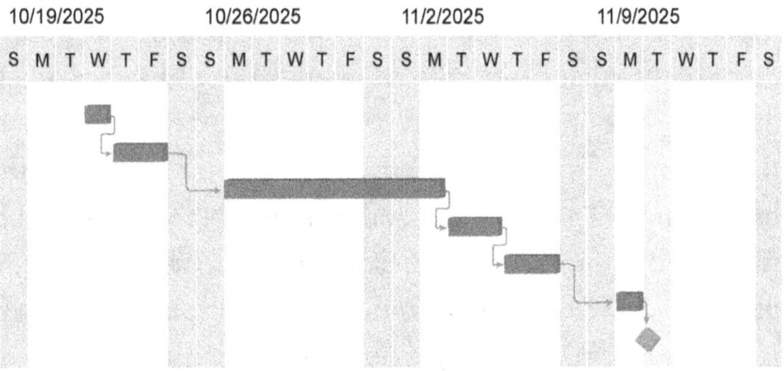

Source: GAO. | GAO-16-89G

Some scheduling software packages include summary activities as an option. Summary activities are grouping elements that are useful for showing the time that activities of lower levels of detail require. Summary activities derive their start and end dates from lower-level activities. Because the work is done at the level of detailed activities, summary activities should never be linked to or from other activities. Figure 4 shows a summary activity, rolling up the time and effort required to complete framing.

---

[4]Detail activities can be defined in different ways in an earned value management (EVM) system. Appendix III provides an overview of schedules and EVM.

Best Practice 1: Capturing All Activities

Figure 4: Summary Activities

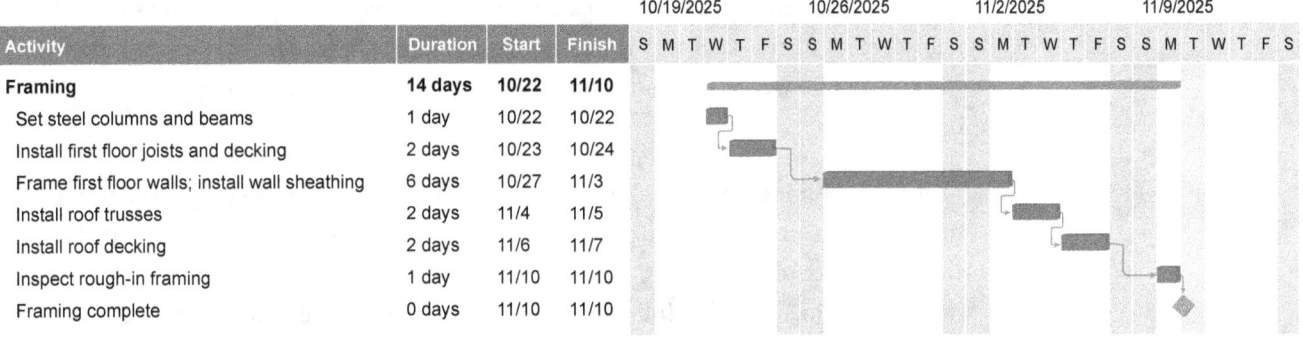

Activity	Duration	Start	Finish
Framing	**14 days**	**10/22**	**11/10**
Set steel columns and beams	1 day	10/22	10/22
Install first floor joists and decking	2 days	10/23	10/24
Frame first floor walls; install wall sheathing	6 days	10/27	11/3
Install roof trusses	2 days	11/4	11/5
Install roof decking	2 days	11/6	11/7
Inspect rough-in framing	1 day	11/10	11/10
Framing complete	0 days	11/10	11/10

Source: GAO. | GAO-16-89G

Level-of-Effort Activities

In addition to detailed work activities and milestone events, other activities in schedules represent effort that has no measurable output and cannot be associated with a physical product or defined deliverable.[5] These level-of-effort (LOE) activities are typically related to management and other oversight that continues until the detailed activities they support have been completed. Progress on LOE activities is based on the passage of time, not the accomplishment of some discrete effort. In schedules, they should be seen as contributing to the comprehensive plan of all work that is to be performed.

LOE is represented by summary activities in certain scheduling software packages, and some schedulers represent LOE effort with detailed activities of estimated long duration. When represented as summary or long-duration detailed activities in a fully networked schedule, LOE activities may inadvertently define the length of a project or program and become critical. LOE activities should never be on the critical path because they do not represent discrete effort. A way to circumvent issues associated with including LOE in a schedule is to represent the effort as an activity that has been designed specifically for that purpose—that is, one that derives its duration from detailed activities. These types of activities, used specifically to represent LOE, are known as "hammock" activities, and they are included in certain scheduling software packages.[6]

CONSTRUCTING AN IMS

The overall size of the IMS depends on many factors, including the complexity of the program and its technical, organizational, and external risks. In addition, the intended

---

[5]In terms of earned value management, a third type of activity in addition to detail work and level-of-effort activities is called apportioned effort. See appendix III for more information.

[6]In the absence of hammock activities, certain techniques allow for the use of LOE-type long-duration activities in a schedule without their interfering with critical path calculations. For example, schedulers may choose to avoid the use of logic links on LOE activities or they may create LOE activities that are one day shorter than the actual planned project length. Because these techniques are used to circumvent the impacts of long-duration activities on traditional critical path calculations, their use and implications should be thoroughly documented in the schedule narrative and basis documents (described later in the guide).

use of the schedule in part dictates its size. That is, a schedule with many short-duration activities may make the schedule unusable for other purposes such as strategic management or risk analysis. The schedule should not be so detailed as to interfere with its use. However, the more complex a program is, the more complex the IMS may become.

Generally speaking, the level of detail in the schedule should reflect the level of information available on the portion of the work that is planned to be accomplished. Both the government and the contractor must define the effort required to complete the program in a way that fully details the entire scope and planned flow of the work. In this manner, the IMS is defined to the level necessary for executing daily work and regularly updating the program. Schedules that are defined at too high a level may disguise risk that is inherent in lower-level activities. In contrast, too much detail in a schedule will make it difficult to manage progress and may convolute the calculation of critical paths.

The IMS ideally takes the form of a single schedule file that includes all activities. However, it may also be a set of separate schedules, perhaps representing the work of separate contractors and government offices, networked together through external links. Regardless of how this is achieved, the IMS schedules must be consistent horizontally and vertically. Horizontal and vertical integration forms the basis of Best Practice 5.

The IMS includes the summary, intermediate, and all detailed schedules. At the highest level, a summary schedule should provide a strategic view of summary activities and milestones necessary to start and complete a program. Decision makers use summary schedules to view overall progress toward key milestones. Summary schedules are roll-ups of lower-level intermediate and detail schedules. The dates of these milestones are automatically calculated through the established network logic between planned activities.

An intermediate schedule includes all information displayed in the summary schedule, as well as key program activities and milestones that show important steps toward high-level milestones. Intermediate schedules may or may not include detailed work activities. For instance, an intermediate schedule may show the interim milestone accomplishments necessary before a major milestone decision or summarized activities related to a specific trade or resource group. A properly defined IMS can facilitate tracking key program milestones such as major program decision points or deliverables. The important program milestones can be summarized along with the specific required activities leading up to the milestone event.

A detailed schedule, the lowest level of schedule, lays out the logically sequenced near-term effort to achieve program milestones. While each successively lower level of schedule shows more detailed date, logic, resource, and progress information, summary, intermediate, and detailed schedules should be integrated in a way such that higher-level schedule data respond dynamically and realistically to progress (or lack of progress) at the lower levels. Delays in lower-level schedules should be immediately rolled up to intermediate and summary schedules. A summary schedule presented to senior management should not display on-time progress and on-time finish dates if the same milestones in lower-level schedules are delayed.

Ideally, the same schedule serves as the summary, intermediate, and detailed schedule by simply rolling up lower levels of effort into summary activities or higher-level WBS elements. When fully integrated, the IMS shows the effect of delayed or accelerated government activities on contractor activities, as well as the opposite. Not every team member needs to digest all the information in the entire schedule. For example, decision makers need strategic overviews, whereas specialist contractors need to see the detail of their particular responsibility. Both sets of information should be available from the same data in the same schedule.

Management should take steps to ensure the accuracy of reported schedule information. In some instances, the government program management office and its contractors might use different scheduling software. However, given the same schedule data, different software products will produce different results because of variations in algorithms and functionality. Attempting to manually resolve incompatible schedules in different software can become time-consuming and expensive. If the use of different software cannot be avoided, the parties should define a process to preserve integrity between the different schedule formats and to verify and validate the converted data whenever the schedules are updated. To ensure integration, milestones need to be defined between the government and the contractor schedules. These milestones are sometimes referred to as "giver/receiver" milestones and one of their purposes is to ensure that integrated schedules reflect the same dates. Giver/receiver milestones are described in more detail in Best Practice 5.

The IMS must include planning for all activities that have to be accomplished for the entire duration of the program. A schedule of planned effort for one block, increment, or contract for a multiyear multiphased program is not a plan sufficient to reliably forecast the finish date for the program. Without such a view, a sound basis does not exist for knowing with any degree of confidence when and how the program will be completed. A comprehensive IMS reflects all activities for a program and recognizes that there can be uncertainties and unknown factors in schedule estimates because of limited data, technical difficulty, inadequate resources, or other factors in the organizational environment.

Uncertainties regarding future activities are incorporated into an IMS in part by the rolling wave process (discussed in Best Practice 3) and through schedule risk analysis (Best Practice 8). Management should verify that all subcontractor schedules are correctly integrated in the IMS with detail appropriate to their risk level. Case study 2 gives an example of high-level program dates unsupported by planned effort.

For this DOD study, we reviewed the most current schedule and cost estimates that supported DOD's February 2012 Milestone B decision, which determined that investment in the Defense Enterprise Accounting and Management System was justified. We found that the schedule used to support the Milestone B decision included the activities to be performed by both the government and the contractor for Releases 1 through 3 of Increment 1. However, the schedule did not reflect activities to be performed for Releases 4 through 6 of Increment 1 or for Releases 1 and 2 of Increment 2. The DEAMS program manager stated that a comprehensive schedule for Increment 1 that included the activities for all six releases would not be completed until mid-2014. The program manager also stated that Increment 2 had not been included because program officials did not know the detailed activities to be performed that far in advance.

To address this issue, the DEAMS program office developed a roadmap—a planning document that briefly outlines the program's key increments and releases and expected milestones for completion—depicting Releases 1 through 6 of Increment 1 and Releases 1 and 2 of Increment 2 with a full deployment date of fiscal year 2017. However, the program office did not provide a schedule that supported the estimated dates in the roadmap.

A schedule incorporates different levels of detail depending on the information available at any point in time. That is, near-term effort is planned in greater detail than long-term effort. Effort beyond the near term that is less well defined is represented within the schedule as long-term planning packages. Planning packages are a summarization of the work to be performed in the distant future with less specificity. By not including all work for all deliverables for both increments and all releases, the DEAMS program could incur difficulties resulting from an incomplete understanding of the plan and what constitutes a successful conclusion for the program. DEAMS program officials provided a draft of the Schedule Management Plan that documented their intent to use a planning package approach when updating the DEAMS schedule in the future.

GAO, *DOD Business Systems Modernization: Air Force Business System Schedule and Cost Estimates*, GAO-14-152 (Washington, D.C.: February 7, 2014).

THE IMS AS A CONSOLIDATION TOOL

Typically, an IMS is constructed to establish logic links and share resources across related projects of a single program. However, nothing requires projects embedded in a master schedule to share links or resources or to relate to one another in the context of an overall program. An IMS can be a useful tool for consolidating multiple project files in a single master file, even if those projects are immaterially related. For example, aggregating individual files in a master schedule is useful for reporting purposes, particularly if the projects are under the purview of a single management organization or a single customer. In this case, the master schedule allows for a concise view of all projects for which the stakeholder is responsible or has an interest. A master schedule of this nature is often referred to as a consolidated schedule or a portfolio schedule, although these terms are often synonymous with IMS.

Case study 3 highlights the usefulness of creating an IMS from individual projects that are within the purview of a single client, share resources, and yet have no logic dependencies between them.

Case Study 3: Consolidated program schedules, from *Arizona Border Surveillance Technology Plan*, GAO-14-368

Under the Secure Border Initiative Network (SBI*net*), the Department of Homeland Security's (DHS) U.S. Customs and Border Protection (CBP) deployed surveillance systems along 53 of the 387 miles of the Arizona border with Mexico. After DHS canceled further SBI*net* procurements, CBP developed the Arizona Border Surveillance Technology Plan (the Plan), which includes a mix of radars, sensors, and cameras to help provide security for the remainder of Arizona's border. GAO was asked to review the status of DHS's efforts to implement the Plan, including the extent to which CBP had developed schedules in accordance with best practices.

DHS had not developed an IMS for scheduling, executing, and tracking the work to implement the Plan and its seven programs. Rather, DHS had used a separate schedule for each individual program to manage the implementation of the Plan. DHS officials stated that an IMS for the overarching Plan was not needed because the Plan contained individual acquisition programs as opposed to a plan consisting of seven integrated programs.

However, collectively, these programs were intended to provide CBP with a combination of surveillance capabilities to assist in achieving situational awareness along the Arizona border with Mexico, as referenced in CBP's planning documents. Moreover, while the programs themselves may have been independent of one another, the Plan's resources were being shared among the programs. DHS officials stated that when they developed schedules for the Plan's programs, they assumed that personnel would be dedicated to work on individual programs and not be shared between programs. However, as DHS initiated and continued work on the Plan's programs, it shared resources such as personnel among the programs, contributing, in part, to delays in the programs.

Further, DHS officials stated that because of resource constraints associated with the initiation of the Plan, the development of two acquisition documents was deferred. In addition, planning and deployment activities for some programs were delayed because of resource-constrained environments and the lack of dedicated contracting officers to plan and execute the programs' source selection and environmental activities.

Developing and maintaining an IMS for the Plan could have allowed DHS insight into current or programmed allocation of resources for all programs as opposed to attempting to resolve resource constraints for each program individually. Because DHS did not have an IMS for the Plan, it was not well positioned to understand how schedule changes in each individual program affected implementation of the overall Plan.

GAO, *Arizona Border Surveillance Technology Plan: Additional Actions Needed to Strengthen Management and Assess Effectiveness*, GAO-14-368 (Washington, D.C.: March 3, 2014).

WORK BREAKDOWN STRUCTURE

A work breakdown structure is the cornerstone of every program because it defines in detail the work necessary to accomplish a program's objectives. For example, a typical WBS reflects the requirements to be accomplished to develop a program, and it provides a basis for identifying resources and activities necessary to produce deliverables. A WBS is also a valuable communication tool between systems engineering, program management, and other functional organizations because it provides a clear picture of what has to be accomplished by decomposing the project scope into finite deliverables. Accordingly, it is an essential element for identifying activities in a program's IMS.

A well-structured WBS helps promote accountability by identifying work products that are independent of one another. It also provides the framework for developing a schedule plan that can easily track technical accomplishments—in terms of resources spent in relation to the plan as well as completion of activities—allowing quick identification of cost and schedule variances.

Figure 5: The WBS and Scheduling

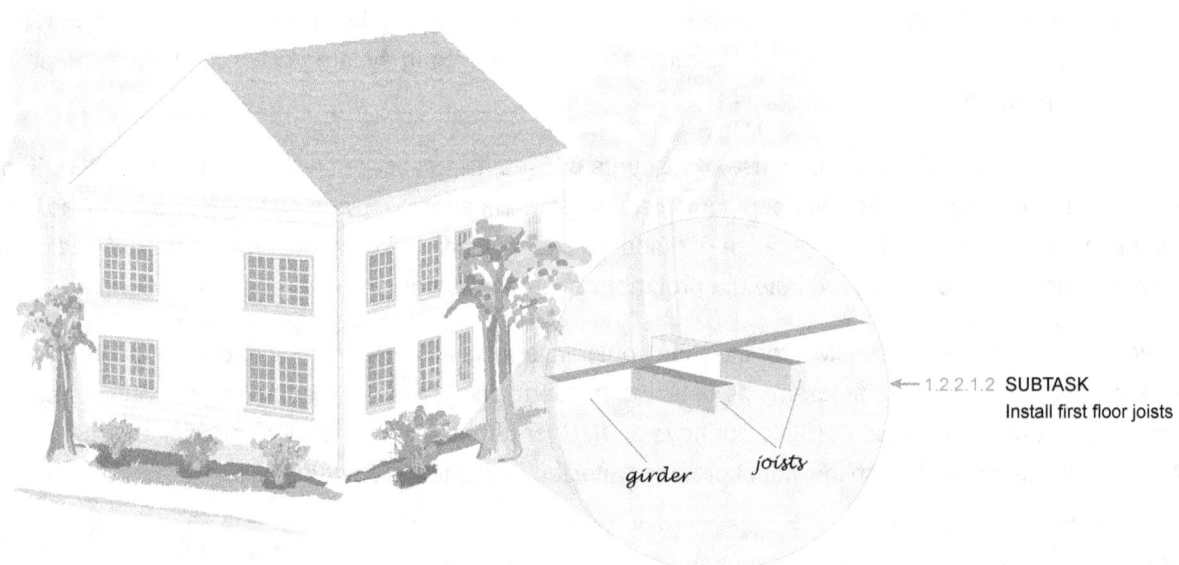

PLANNING

0.0 PROGRAM Home
1.0 PROJECT House
1.1 PHASE Start-up
1.2 PHASE Construction
1.2.1 TASK Foundation work
1.2.2 TASK House construction

SCHEDULING

1.2.2.1 WORK PACKAGE Framing
1.2.2.1.1 SUBTASK Set steel columns
1.2.2.1.2 SUBTASK Install first floor joists

1.2.2.1.2 SUBTASK Install first floor joists

girder   joists

Source: GAO, adapted from © 2010 Paul D. Giammalvo. | GAO-16-89G

Best Practice 1: Capturing All Activities

A WBS deconstructs a program's end product into successively greater levels of detail until the work is subdivided to a level suitable for management control. By breaking work down into smaller elements, management can more easily plan and schedule the program's activities and assign responsibility for the work. It is also essential for establishing a reliable schedule baseline. Establishing a product-oriented WBS such as the one in figure 5 is a best practice because it allows a program to track cost and schedule by defined deliverables, such as a hardware or software component. This allows program managers to identify more precisely which components are causing cost or schedule overruns and to more effectively mitigate overruns by manipulating their root cause.

The WBS is the basis of the program schedule and defines what is required as a deliverable. Scheduling activities addresses how the program is going to produce and deliver what the WBS describes. When properly planned, the schedule reflects the WBS and therefore defines the activities necessary to produce and deliver the lowest-level deliverable. In essence, the schedule is a set of instructions on how the program intends to execute. Every activity within the schedule should be traceable to an appropriate WBS element, and every WBS element must have at least one associated activity that is necessary to complete that element clearly identified within the schedule. Aligning the schedule to the program WBS will help ensure that the total scope of work is accounted for within the schedule.

In the schedule, the WBS elements are linked to one another through the activities' logical relationships and they lead to the end product or final delivery. The WBS progressively deconstructs the deliverables of the entire effort through lower-level elements.

It is important that the WBS is comprehensive enough to represent the entire program in detail sufficient to manage the size, complexity, and risk associated with the program. There should be only one WBS for each program, and it should match the WBS used for the cost estimate and schedule so that actual costs can be fed back into the estimate with a correlation between the cost estimate and schedule. A well-developed WBS is essential to the success of all acquisition programs.[7]

In addition, the WBS must have an associated WBS dictionary that clearly defines the scope of each individual WBS element and, therefore, the scope of related schedule activities. Activities that are not assigned to WBS elements or are assigned to undefined WBS elements reflect unassigned or undefined scope. In this manner, the WBS dictionary helps clarify the boundaries between different WBS elements.

---

[7]See the GAO *Cost Estimating and Assessment Guide* (GAO-09-3SP) for numerous examples of standard work breakdown structures for, among others, surface, sea, and air transportation systems; military systems; communications systems; and systems for construction and utilities. For example, DOD has identified, for each defense system, a standard combination of hardware and software that defines the end product for that system. In its WBS standard, DOD defines and describes the WBS, provides instructions on how to develop one, and defines specific defense items (*Department of Defense Standard Practice: Work Breakdown Structures for Defense Materiel Items,* MIL-STD-881C (Washington, D.C.: October 3, 2011).

Statement of Work

The statement of work (SOW) defines, either directly or by reference to other documents, performance requirements for a contractor's effort. The SOW defines the scope of the contract—that is, it specifies the work to be done in developing the goods or services to be provided by a contractor. Activities within a contractor's project schedule should be directly traceable to the program's SOW.[8]

The hierarchical nature of the WBS ensures that the entire SOW accounts for the detailed technical activities and, when completed, facilitates communication between the customer and supplier on cost, schedule, resource requirements, technical information, and the progress of the work.

ACTIVITY NAMES

A consistent convention for naming work activities should be established early in a program and carried through its completion. Names for all activities, including summary, milestone, and detailed activities, should be unique and as descriptive as necessary to facilitate communication between all team members. Descriptive activity names ensure that decision makers, managers, activity managers, task workers, and auditors know what scope of work is required for each activity.

Because activities are essentially instructions for someone to carry out, activity names should be phrased in the present tense with verb-noun combinations, such as "review basis of estimate," "test level 4 equipment," or "install and test workstations, room 714." Milestone descriptive names should be related to an event or a deliverable, such as "Milestone B decision" or "level 4 test results report delivered." In all cases, descriptive names should be unambiguous and should identify their associated product without the need to review high-level summary activity or preceding activity names. In the example in figure 6, it is not clear which products the detailed inspection activities are associated with if their respective summary activities and preceding activities are not included in the filtered view.

---

[8]We use *SOW* as a generic term to represent a document that sets forth the scope of a contract. A statement of objective (SOO) and performance work statement (PWS) may also be used in the contractual process to establish desired service outcomes and performance standards.

Figure 6: Redundant Activity Names

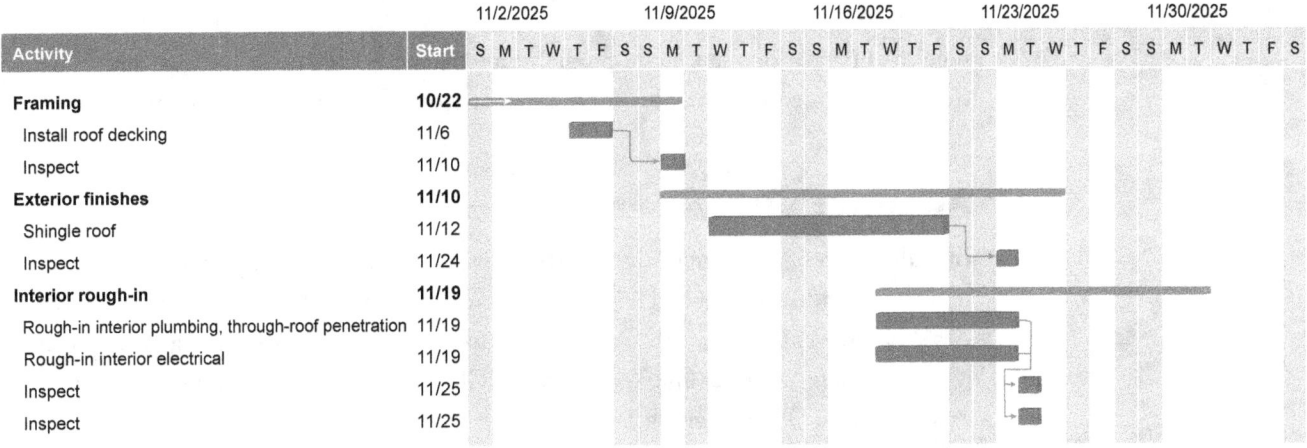

Source: GAO. | GAO-16-89G

Repetitive naming of activities that are not associated with specific products or phases makes communication difficult between teams, particularly between team members responsible for updating and integrating multiple schedules. However, in figure 7, activity names include their respective products, which are also identified for clarity in their summary activity.

Figure 7: Unique Activity Names

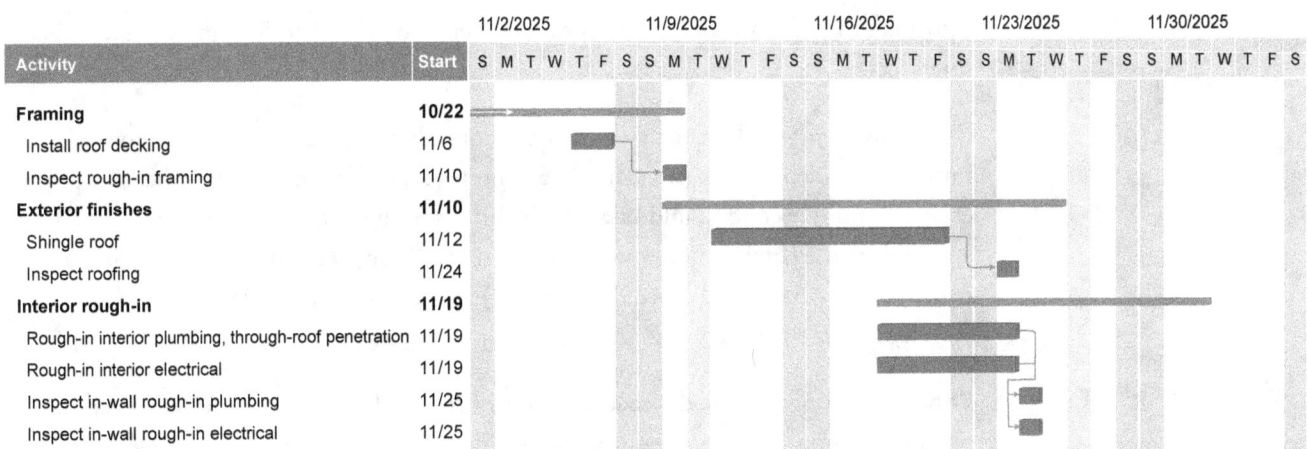

Source: GAO. | GAO-16-89G

Individual activities in figure 7 are easier to identify, especially when filters and other analyses are performed on the schedule data and the activities are taken out of the summary-indented activity context. For example, filtering the schedule data to display only critical path activities will not be helpful if activity names are redundant or incompletely identify the work being executed. Communicating to management that "inspection" is a critical path activity is not useful unless management knows which inspection is meant. Communicating that "drywall screw inspection" is a critical path activity conveys usable information and obviates the need to note the summary or predecessor activity. Unique activity names are essential if a schedule is to produce reliable information at the summary, intermediate, and detailed levels.

Finally, abbreviations specific to programs and agencies should be minimal and, if used, should be defined in either the WBS dictionary or the schedule basis document.[9]

ACTIVITY CODES

In addition to having descriptive names, activities within the schedule must be consistent with key documents and other information through activity or task codes. Activities should be consistent with such information as the related WBS element, contractor, location, phase, contract line item number (CLIN), work package number, control account, and SOW paragraph as applicable.[10] Consistent coding can help ensure the vertical integration of the summary and detailed work schedules. These data should be stored in custom fields rather than appended to activity names.

It is also helpful to include references to the organizational breakdown structure (OBS). The OBS reflects the responsible organization of the project and describes the hierarchy that will provide resources to perform the work identified in the WBS. If work is delayed by an issue within an organization, OBS codes within the schedule allow planners to quickly identify the activities that are affected. In addition, every activity should be identified with an activity owner. The activity owner is the person responsible for understanding the scope of work and objectives, defining the entrance and exit criteria, providing the schedule updates to the scheduler, and understanding any variances that might occur.

Each activity should be associated with a unique alpha-numeric code that allows it to be immediately identified, particularly in a complex schedule with many embedded projects. This unique code should preserve the hierarchical relationship between the activity and its parent elements. For example, the activity "Software Unit 1A test" could be mapped to the code "ADZT125" to represent the Alpha project (A), development phase (D), Zeta prime contractor (Z), testing activity (T), and activity ID number (125).

Other commonly used codes that are mapped to schedule activities, either placed in separate columns or embedded in a unique identifier, identify the responsible owner of the activity and the related integrated product team (IPT). Codes such as phase, department, area, system, and step also greatly facilitate the filtering and organization of schedule data for reporting, metric analysis, and auditing purposes. LOE activities should

---

[9] A schedule basis document and its recommended content are detailed in Best Practice 10.

[10] In addition, some agencies, particularly DOD, may choose to align the IMS to an integrated master plan (IMP). The IMP is an event-based plan that allows the tracking of accomplishments for each event, as well as completion criteria for each accomplishment. The IMS flows directly from the IMP and includes the detailed activities necessary to support the criteria. The IMS is traceable to all IMP events, accomplishments, and criteria. When used, the IMP is typically contractually binding.

be identified as such in the schedule, and government and contractor efforts should be clearly delineated. Any custom text fields or coding scheme within the schedule should be defined in the schedule basis document. Case study 4 shows how activities in a detailed schedule can be quickly sorted, filtered, or categorized by means of codes.

Case Study 4: Successfully Coding Detailed Work, from *VA Construction*, GAO-10-189

The Department of Veterans Affairs (VA) required by contract that the schedule for the expansion of the medical centers in Cleveland, Ohio, include approximately 2,500 activities in order to sufficiently detail the level of work required. The actual schedule contained 2,725 activities—approximately 75 detail activities per milestone. Each activity was mapped to an activity identification number, building area, and work trade, which allowed the scheduler to quickly filter the schedule by type of work or subcontractor. Additionally, the VA Office of Construction and Facilities Management reviewed the schedule for completeness to ensure that all necessary activities and milestones were included.

GAO, *VA Construction: VA Is Working to Improve Initial Project Cost Estimates, but Should Analyze Cost and Schedule Risks*, GAO-10-189 (Washington, D.C.: December 14, 2009).

Best Practices Checklist: Capturing All Activities

- A WBS is the cornerstone of the program schedule. Its elements are linked to one another with logical relationships and lead to the end product or final delivery. The schedule clearly reflects the WBS and defines the activities necessary to produce and deliver each product.

- The schedule reflects all effort (steps, events, work required, and outcomes) to accomplish the deliverables described in the program's work breakdown structure.

- The IMS includes planning for all activities that have to be accomplished for the entire duration of the program, including all blocks, increments, phases, and the like.

- The IMS includes the summary and intermediate and all detailed schedules. The same schedule serves as the summary, intermediate, and detailed schedule by simply rolling up lower levels of effort into summary activities or higher-level WBS elements.

- The government-owned detailed schedule includes all activities the government, its contractors, and others must perform to complete the work, including receipt of government-furnished equipment or information, deliverables, or services from other programs.

- The schedule contains primarily detail activities, and milestones are not used to represent work.

- If the government program management office and its contractor use different scheduling software packages, a process is defined to preserve integrity between the different schedule formats, and the converted data are verified and validated when the schedules are updated.

- Level-of-effort (LOE) activities represent effort that has no measurable output and cannot be associated with a physical product or defined deliverable.

- Activity names contain noun-verb combinations, are descriptive, and are clear enough to identify their associated product without the need to review high-level summary or predecessor activity names.

- Activities within the schedule are easily traced to key documents and other information through activity or task codes.

BEST PRACTICE 2

Sequencing All Activities

Best Practice 2: The schedule should be planned so that critical program dates can be met. To do this, activities must be logically sequenced and linked—that is, listed in the order in which they are to be carried out and joined with logic. In particular, a predecessor activity must start or finish before its successor. Date constraints and lags should be minimized and justified. This helps ensure that the interdependence of activities that collectively lead to the completion of activities or milestones can be established and used to guide work and measure progress.

Once established, a schedule network can forecast reliably—in light of the best information available at that point in time—the start and finish dates of future activities and key events based on the status of completed and in-progress activities. Dates are forecast with some realism by planning work effort in sequences of activities that logically relate portions of effort to one another. The schedule network is a model of all ongoing and future effort related to the program; it establishes not only the order of activities that must be accomplished but also the earliest and latest dates on which those activities can be started and finished to complete the program on time.

The purpose of sequencing activities is to develop a networked schedule that is a predictive model of how it is intended that the program be executed. By establishing the network logic, the schedule can predict the effect on the program's planned finish date of, among other things, misallocated resources, delayed activities, external events, scope changes, unrealistic deadlines, and the effect of risk events.

The ability of a schedule to forecast the start and finish dates of activities and key events reliably is directly related to the complexity and completeness of the schedule network. The reliability of the schedule is in turn related to management's ability to use the schedule to direct the assignment of resources and perform the correct sequence of activities. Activities are related through different types of sequential and parallel predecessor and successor logic.

The more complex a program is, the more complex the schedule may become. However, it is essential that the schedule be as straightforward as possible so that management has a clear indication of the path forward and the necessary resources needed to accomplish activities on time. Convoluted logic techniques such as date constraints, lags, and misused logic links should be eliminated. If they are used, they should be employed judiciously and justified in the schedule's documentation.

Once all dependencies are accounted for, the schedule should be presented for the review and approval of management and the persons responsible for performing the work. Major handoffs between groups should be discussed and agreed on to ensure that the schedule correctly models what they expect to happen. This will help everyone see the big picture needed to complete the entire program.

PREDECESSOR AND SUCCESSOR LOGIC

Activities that are logically related within a schedule network are referred to as predecessors and successors. A predecessor activity must start or finish before its successor. The purpose of a logical relationship, or dependency, is to depict the sequence in which activities occur. Such relationships state when activities are planned to start and finish in relation to the start and finish of other activities. A logic relationship therefore models the effect of an on-time, delayed, or accelerated activity on subsequent activities. Relationships between activities can be internal—that is, within a particular schedule—or external—that is, between schedules.

Logical relationships between activities identify whether they are to be accomplished in sequence or in parallel. A sequence of activities is a serial path along which one activity is completed after another. Activities can also be accomplished in parallel or concurrently. A logic relationship linking a predecessor and successor activity can take one of three forms: finish-to-start, start-to-start, and finish-to-finish.

A finish-to-start (F–S) relationship is the most straightforward logical link between a predecessor and successor. In it, a successor activity cannot start until the predecessor activity finishes, creating a simple sequence of planned effort. This logical relationship is the default in most scheduling programs. In figure 8, the installation of roof decking cannot begin until the installation of roof trusses finishes. Note that the installation of the roof decking does not necessarily need to start once the installation of the roof trusses finishes, but it cannot start until the trusses are installed. The "install roof decking" activity may have other predecessors that push its start date further into the future.

Figure 8: A Finish-to-Start Relationship

Activity	Duration	Start	Finish
Install roof trusses	2 days	11/4	11/5
Install roof decking	2 days	11/6	11/7

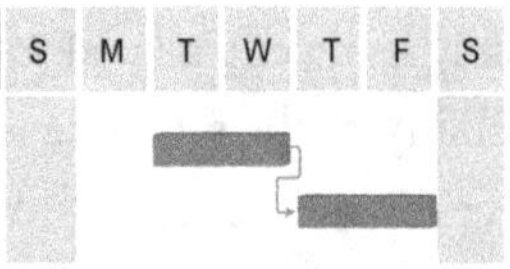

Source: GAO. | GAO-16-89G

A start-to-start (S–S) relationship dictates that a successor activity cannot start until the predecessor activity starts. In the example in figure 9, the application of wall finishes cannot start until the application of drywall texture starts. The S–S relationship does not dictate that wall finishing must start at the same time that drywall texturing starts, but it does indicate that it cannot start until drywall texturing starts. The "apply wall finishes" activity may have other predecessors that push its start date further into the future.

Figure 9: A Start-to-Start Relationship

Activity	Duration	Start	Finish	S	M	T	W	T	F	S
Apply drywall texture	3 days	12/17	12/19							
Apply wall finishes (stain and paint)	3 days	12/17	12/19							

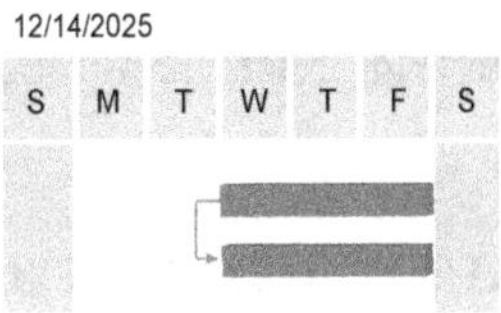

Source: GAO. | GAO-16-89G

A finish-to-finish (F–F) relationship dictates that a successor activity cannot finish until the predecessor activity finishes. In figure 10, final grading cannot finish until forming and pouring the driveway has finished. Again, the F–F relationship does not dictate that landscaping must finish at the same time as pouring the driveway finishes, but it cannot finish until the pouring of the driveway finishes. The final landscaping activity may have other predecessors that determine its finish date, or it may finish later than the driveway pouring activity.

Figure 10: A Finish-to-Finish Relationship

Activity	Duration	Start	Finish	S	M	T	W	T	F	S
Form and pour driveway	2 days	12/1	12/2							
Finish grade property	1 day	12/2	12/2							

Source: GAO. | GAO-16-89G

The start-to-finish (S–F) link is a theoretical, fourth combination of logical links between predecessor and successor. It has the bizarre effect of directing a successor activity not to finish until its predecessor activity starts, in effect reversing the expected flow of sequence logic. Its use is widely discouraged because it is counterintuitive and it overcomplicates schedule network logic. Examples of activity sequences used to justify the existence of an S–F relationship can usually be rewritten in simple F–S logic by either subdividing activities or finding more appropriate F–S predecessors within the network.

The majority of relationships within a detailed schedule should be finish-to-start. Finish-to-start relationships are intuitive because most work is accomplished serially in that order. Moreover, F–S relationships are easy to trace within a schedule network and clearly indicate to management which activities must finish before others begin and which activities may not begin until others have been completed. F–S relationships are implemented most easily where work is broken down to small elements.

Start-to-start and finish-to-finish relationships, in contrast, imply parallel or concurrent work. S–S and F–F relationships represent a valid technique for modeling the overlapping of activities. As such, they may be more predominant in schedules whose detail has not yet evolved. However, an overabundance of these relationships in detail schedules may suggest an overly optimistic or unrealistic schedule or shortcuts that have been taken in modeling activities and logic. Particularly in a detailed schedule, their overuse may impair the usefulness of the schedule by, for example, complicating the identification of the critical path. S–S and F–F relationships are also prone to producing unintentional "dangling" relationship logic, an error that we describe later in this best practice.

EARLY AND LATE DATES

Once the logic that creates a network has been established, the scheduling software can calculate a set of start and finish times for each activity, given the relationship logic and the estimated duration of each activity. Ideally, only one date should be entered into the scheduling software—the program start date. All other dates are calculated by the network logic. Network logic calculates activity dates that define both when an activity may start and finish and when an activity must start and finish to meet a specified program completion date. These are known as early and late dates, respectively.

Each activity has an early start date and an early finish date. Unless otherwise stated in this guide, "start" and "finish" refer to an activity's early start and early finish dates.

- Early start defines the earliest time when an activity may start.
- Early finish defines the earliest time when an activity may finish.

The early start and early finish dates for each activity are calculated by the "forward pass" method. The forward pass determines the early start and early finish times for each activity by adding durations successively through the network, starting at day one. The forward pass will derive the total time required for the entire program by calculating the longest continuous path through the network—that is, the sequence of activities that determines the length of the program, typically known as the critical path. Managing the critical path is the foundation of Best Practice 6.

Each activity also has a late start date and a late finish date.

- Late start defines the latest time when an activity must start in order for the program to be completed on time.
- Late finish defines the latest time when an activity must finish in order for the program to be completed on time.

The late start and late finish for each activity are calculated by the "backward pass" method—the opposite of the forward pass. Whereas the forward pass determines early finishes by adding durations to early starts, the backward pass determines late starts by subtracting durations from late finishes.

The difference between an activity's early and late dates is known as total float or slack. Total float is an essential output of critical path method scheduling, and its proper management is the foundation of Best Practice 7.

In figure 11, setting steel columns and beams may begin as early as November 12 (early start) and finish as early as November 21 (early finish). However, for the house construction project to be completed on time, setting steel columns and beams must begin by November 13. The activity therefore has 1 day of total float: the difference between its early start date of November 12 and its late start date of November 13. The blue box in figure 11 represents the early start and early finish dates, and the red box represents the late start and late finish dates.

Figure 11: Early and Late Dates

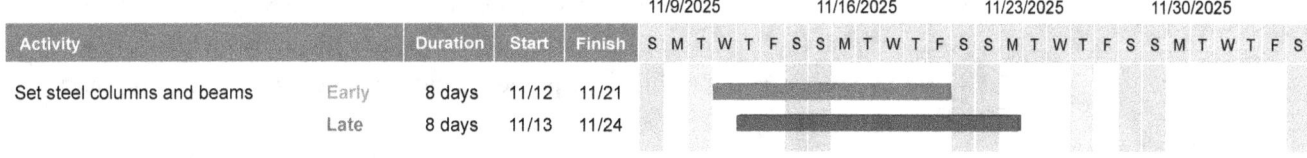

Activity		Duration	Start	Finish	11/9/2025	11/16/2025	11/23/2025	11/30/2025
					S M T W T F S	S M T W T F S	S M T W T F S	S M T W T F S
Set steel columns and beams	Early	8 days	11/12	11/21				
	Late	8 days	11/13	11/24				

Source: GAO. | GAO-16-89G

Appendix IV gives a detailed example of how to calculate early and late dates using the forward and backward passes.

INCOMPLETE AND DANGLING LOGIC

A logic relationship dictates the effect of an on-time, delayed, or accelerated activity on subsequent activities. Any missing or incorrect logic relationship is potentially damaging to the entire network. Complete network logic between all activities is essential if the schedule is to correctly forecast the start and end dates of activities within the plan.

As a general rule, every activity within the schedule should have at least one predecessor and at least one successor. The two natural exceptions to this rule are the program start milestone, which has no predecessor, and the program finish milestone, which has no successor. Other activities or milestones within the schedule may have no predecessor or successor links when they represent schedule inputs or outputs. For example, a milestone may represent the handing off of some interim product to an external partner by the program office that therefore has no successor relationship within the schedule. However, any activity that is missing predecessor or successor logic must be clearly justified in the schedule documentation.

Even if an activity has predecessor and successor logic relationships, incorrect or incomplete logic can arise. Networks should be assessed for circular logic—that is, logic that forces two activities to be dependent on each other. Circular logic creates an endless loop of work: an activity cannot have its successor also be its predecessor. In addition, the network should be clear of redundant logic. Redundant logic represents unnecessary logic links between activities. For example, a sequence of activities A, B, and C with a series of finish-to-start logic has no need for an additional F–S logic link between A and C.

Activities with S–S or F–F relationships should be checked for two types of "dangling" or "hanging" logic. Dangling logic is scheduling logic with an improper tie to an activity's start or end date. Each activity's start date—other than the start milestone—should be driven by a predecessor activity, and each activity's finish date—other than the finish milestone—must drive a successor activity's start or finish. Dangling logic, a form of incomplete logic, can interfere with the valid forecasting of scheduled activities.

The first type of dangling logic occurs when an activity has a predecessor and a successor but its start date is not properly tied to logic. In other words, no preceding activity within the schedule is determining the start date of the activity, with either its start (S–S) or finish (F–S). Figure 12 shows a sequence of activities—rough grade property, form and pour driveway, finish grade property, and plant trees and shrubs and install final landscaping.

Figure 12: Start-Date Dangling Logic

Activity	Duration	Start	Finish
Rough grade property	1 day	11/28	11/28
Form and pour driveway	2 days	12/1	12/2
Finish grade property	1 day	12/2	12/2
Plant trees and shrubs, and finalize landscaping	3 days	12/3	12/5

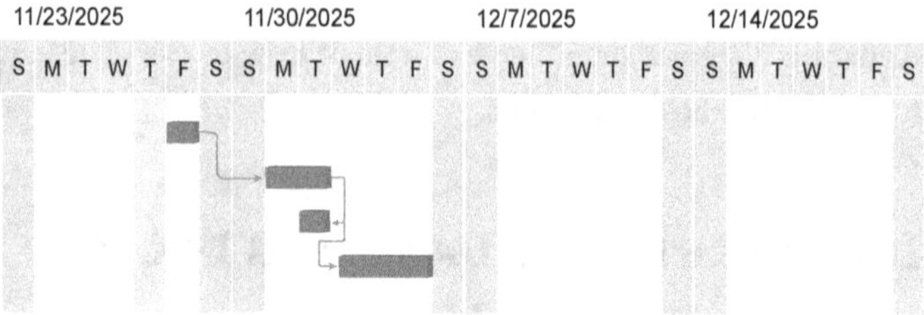

Source: GAO. | GAO-16-89G

"Form and pour driveway" has an F–S relationship to its predecessor activity, "rough grade property," and an F–F relationship with its successor, "finish grade property." "Finish grade property" in turn has an F–S relationship with the "plant trees and shrubs and install final landscaping" activity. Notice that the finish date of "finish grade property" is determined by the predecessor relationship to the finish of "form and pour driveway"—that is, final grading of the property cannot be finished until the driveway is formed and poured. The problem here is that the start date of "finish grade property" is not determined by any relationship; it is determined simply by the estimated time it will take to grade the property and its finish date. Moreover, if "form and pour driveway" finishes but "finish grade property" runs longer than planned, the only resolution—according to the original plan—is for "finish grade property" to start earlier.

This is a logical solution but practically impossible, because it may have already started before it is determined that it takes longer than planned. To correct the dangling logical error, "finish grade property" should have at least one F–S or S–S predecessor link alongside the F–F predecessor relationship to determine its start date. This requirement might make the scheduler find a different predecessor or break "form and pour driveway" into two activities. One solution is shown in figure 13. A link has been added from "rough grade property" to "finish grade property," establishing the logic that landscapers cannot finish grade until the rough grading is complete. According to the new logic model, "finish grade property" cannot start sooner than Monday, December 1.

Figure 13: Start-Date Dangling Logic Corrected with Predecessor Logic

Activity	Duration	Start	Finish	11/23 2025	11/30/2025	12/7/2025	12/14/2025
				S M T W T F S	S M T W T F S	S M T W T F S	S M T W T F S
Rough grade property	1 day	11/28	11/28				
Form and pour driveway	2 days	12/1	12/2				
Finish grade property	1 day	12/2	12/2				
Plant trees and shrubs and finalize landscaping	3 days	12/3	12/5				

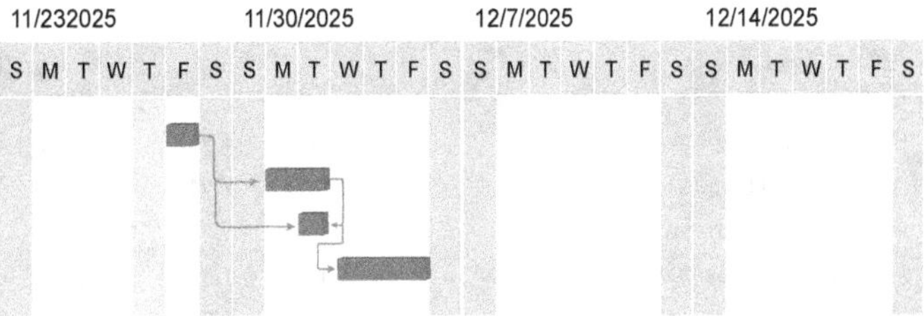

Source: GAO. | GAO-16-89G

The second type of dangling logic is similar to the first but involves an activity's finish date. Figure 14 shows a sequence of activities: finish drywall, apply drywall texture, apply wall finishes, and install trim. "Apply drywall texture" has an F–S predecessor link to "finish drywall" and an S–S successor relationship to "apply wall finishes."

Figure 14: Finish-Date Dangling Logic

Activity	Duration	Start	Finish	12/7/2025	12/14/2025	12/21/2025	12/28/2025
				S M T W T F S	S M T W T F S	S M T W T F S	S M T W T F S
Finish drywall (tape and mud)	5 days	12/10	12/16				
Apply drywall texture	3 days	12/17	12/19				
Apply wall finishes (stain and paint)	3 days	12/17	12/19				
Install interior door, window, baseboard trim	6 days	12/22	12/30				

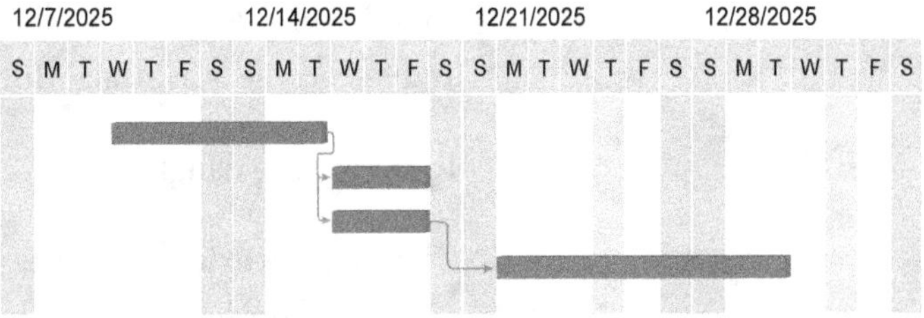

Source: GAO. | GAO-16-89G

In this example, "apply drywall texture" starts once "finish drywall" completes. Once "apply drywall texture" starts, "apply wall finishes" can start. Note, however, that while "apply drywall texture" has a successor, its finish date is not related to any subsequent activity. In other words, "apply drywall texture" can continue indefinitely with no adverse effect on subsequent activities, until it affects the owner's occupation date—to which it is implicitly linked by an F–S relationship by the scheduling software. To correct the dangling logical error, the "apply drywall texture" activity should have at least one F–S or F–F successor link alongside the S–S successor relationship for its finish date to affect downstream activities. Figure 15 shows one possible correction to the dangling logic. An F–S predecessor has been added from "apply drywall texture" to "install tile in bathroom and kitchen." Now, according to the logic model, "apply drywall texture" cannot be delayed without also delaying "install tile in bathroom and kitchen."

Figure 15: Finish-Date Dangling Logic Corrected with Successor Logic

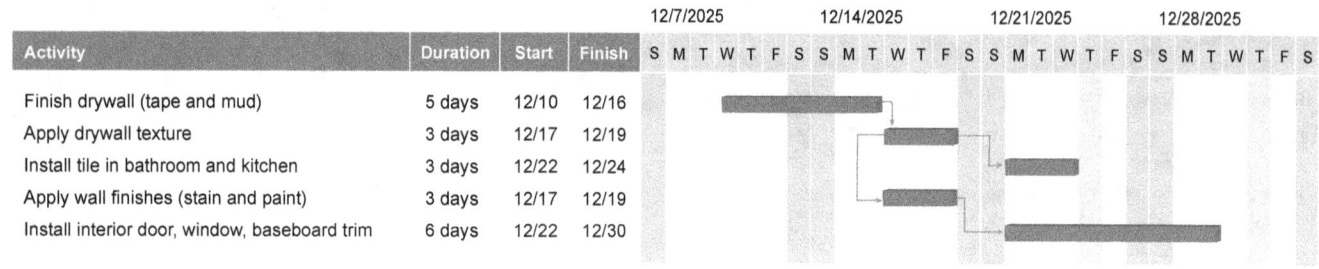

Activity	Duration	Start	Finish
Finish drywall (tape and mud)	5 days	12/10	12/16
Apply drywall texture	3 days	12/17	12/19
Install tile in bathroom and kitchen	3 days	12/22	12/24
Apply wall finishes (stain and paint)	3 days	12/17	12/19
Install interior door, window, baseboard trim	6 days	12/22	12/30

Source: GAO. | GAO-16-89G

Finally, note that dangling logic is far more dangerous for detail activities than milestones. Milestones have the same start and finish dates, so from a practical standpoint, delaying or accelerating a milestone within the schedule would still affect successor activities, even with start or finish date dangling logic. In fact, there is little reason to sequence milestones with any logic other than F–S, because milestones simply indicate a point in time. Regardless, all dangling logic should be corrected to ensure that the logic in the schedule is as straightforward and intuitive as possible.

To summarize dangling logic checks:

- Any activity with an F–F predecessor link should also have at least one F–S or S–S predecessor link. If nothing is driving the start date of the activity, then why not start the activity earlier?
- Any activity with an S–S successor link should also have at least one F–S or F–F successor link. If the finish date of the activity is not driving the start of another activity, then why finish the activity?

Complete schedule logic that addresses the logical relationships between predecessor and successor activities is important. The analyst needs to be confident that the schedule will automatically calculate the correct dates when the activity durations change.

SUMMARY LOGIC

Certain scheduling software packages include summary activities as an option. If used, summary activities should not have logic relationships because their start and finish dates are derived from lower-level activities. Therefore, there is no need for logic relationships on a summary activity in a properly networked schedule.

Summary logic hinders vertical traceability. For example, if the start or finish dates of summary activities are not derived by the planned or actual dates of lower-level activities, then the dates of summary activities may be misrepresented in higher-level versions of the schedule.[11] In addition, tracing logic through summary links does not impart to management the sequence in which lower-level activities should be carried out. Figure

---

[11]Vertical traceability is discussed in detail in Best Practice 5.

16 gives an example of a linked summary activity. The summary activity "Certificate of occupancy" defines the start date of "perform final HVAC inspection" rather than the activity's start date defining the summary start date. Moreover, because the summary logic masks actual work effort relationships, it may not be clear to management that "perform final HVAC inspection," "perform final electrical inspection," and "perform final plumbing inspection" all depend in part on "interior finishes complete."

Figure 16: A Linked Summary Activity

Activity	Duration	Start	Finish
Interior finishes complete	0 days	1/23	1/23
Certificate of occupancy	**9 days**	**1/26**	**2/5**
Perform final HVAC inspection	1 day	1/26	1/26
Perform final electrical inspection	1 day	1/27	1/27
Perform final plumbing inspection	1 day	1/28	1/28
Apply for certificate of occupancy	1 day	1/29	1/29
Approval period for certificate of occupancy	5 days	1/30	2/5

Source: GAO. | GAO-16-89G

Case study 5 provides an example of how using summary links may initially make scheduling activities easier but eventually convolutes the network logic.

Case Study 5: Summary Logic, from *DOD Business Transformation*, GAO-11-53

GAO's analysis of the General Fund Enterprise Business System schedule found that 50 summary activities (12 percent of remaining summary activities) had predecessor links. Program Management Office schedulers used these summary links rather than linking predecessors to their numerous lower-level activities. Because many of the lower-level activities began on the same date, this made updating the schedule simpler: an updated start date for the summary activity forced that same date on all the unlinked lower-level activities.

Despite making updating easier, the technique is not considered a best practice. First, summary activities do not represent work and are used simply as grouping elements. They should take their start and finish dates from lower-level activities; they should not dictate the start or finish of lower-level activities. Second, linking summary activities obfuscates the logic of the schedule. That is, tracing logic through summary links does not impart to management the sequence in which lower-level activities should be carried out.

GAO, *DOD Business Transformation: Improved Management Oversight of Business System Modernization Efforts Needed,* GAO-11-53 (Washington, D.C.: October 7, 2010).

DATE CONSTRAINTS

Ideally, relationship logic, the durations of activities, and resource availability determine the planned early and late start dates of all activities within the schedule network. However, in some cases it may be necessary to override the calculated start or finish dates of activities by imposing calendar restrictions on when an activity can begin or end. Such restrictions are referred to as date constraints. Constraints can be placed on an activity's start or finish date and can limit the movement of an activity to the past or future or both. Because constraints override network logic and restrict how planned dates respond to actual accomplished effort or resource availability, they should be used only when necessary and only if they are justified in the schedule documentation. Generally, constraints are used to demonstrate an external event's effect on the schedule. For example, constraints may be used to show the expected availability of a production line.

"Not earlier than" constraints affect the forward pass of the schedule and thus may delay a program by pushing some activities' start dates later than their predecessors dictate. These types of constraints are also known as "past-limiting" in that they prevent activities from starting or finishing earlier than planned but allow them to slip into the future if predecessor activities are delayed.[12]

- Start no earlier than (SNET): schedules an activity to start on or after a certain date even if its predecessors start or finish earlier. That is, SNET constraints prevent an activity from beginning before a certain date. SNET constraints are also called start on or after constraints.
- Finish no earlier than (FNET): schedules an activity to finish on or after a certain date. That is, FNET constraints prevent an activity from finishing before a certain date. FNET constraints are also called finish on or after constraints.

"Not later than" constraints affect the backward pass of the schedule and thus may unrealistically accelerate a project. These types of constraints are also known as "future-limiting" in that they prevent activities from starting or finishing later than planned but allow them to be accomplished earlier.

- Start no later than (SNLT): schedules an activity to start on or before a certain date. That is, SNLT constraints prevent an activity from starting any later than a certain date. SNLT constraints are also called start on or before constraints.
- Finish no later than (FNLT): schedules an activity to finish on or before a certain date. That is, FNLT constraints prevent an activity from finishing after a certain date. FNLT constraints are also called finish on or before constraints.

"Must" or "mandatory" constraints affect both the forward pass and the backward pass

---

[12]We use date constraint names consistent with September 2013 DOD data exchange instructions supplementing the United Nations Center for Trade Facilitation and E-business (UN/CEFACT) XML schema 09B. These names may differ depending on the scheduling software used. See appendix V for more information and a mapping of common date constraint names.

of a schedule, forcing activities to occur on dates regardless of network logic. These types of constraints prevent activities from starting or finishing on any day other than the date assigned.

- Must start on (MSON): schedules an activity to start on a certain date. That is, MSON constraints prevent an activity from starting any earlier or later than a certain date, thereby overriding network logic. MSON constraints are also called mandatory start constraints.
- Must finish on (MFON): schedules an activity to finish on a certain date. That is, MFON constraints prevent an activity from finishing any earlier or later than a certain date, thereby overriding network logic. MFON constraints are also called mandatory finish constraints.

Date constraints are often categorized as either soft (also referred to as moderate or one-sided) or hard (also referred to as inflexible), depending on how they restrict the ability of the activity to accelerate or slip according to the established network logic.[13] Soft constraints include SNET and FNET constraints. These are called soft because while they restrict an activity from starting or finishing early, depending on network logic, they allow the activity to start or finish later than planned. In this respect, these constraints allow delays to permeate the schedule and, given available float, possibly affect the program's end date.

Hard constraints include SNLT, FNLT, MSON, and MFON constraints. SNLT and FNLT constraints prevent activities from starting or finishing later than planned, essentially restricting the ability of any predecessor delays to affect their start and finish dates. While these types of constraint allow activities to start and finish earlier than planned, the acceleration of activities is not usually as big a concern to program management as the delay of activities.

Mandatory start and finish constraints are the most rigid because they do not allow an activity to either take advantage of time savings by predecessor activities or slip in response to delayed predecessors or longer-than-scheduled durations. By setting the early and late dates of an activity equal to each other, a mandatory start or finish constraint immediately eliminates all float associated with the activity and renders activities static in time; successors might start on the next day, even though unconstrained logic would not permit it.

---

[13]Our definitions of hard and soft constraints assume that a schedule is formulated under forward scheduling. That is, the project start date is firm and all activities are scheduled to occur as soon as possible to determine the project finish date. Under backward scheduling, the project end date is considered firm and all activities are scheduled as late as possible (ALAP) to determine the project start date. Whether the plan is forward or backward scheduled affects how restrictive certain constraints are. For example, under backward scheduling, a finish no earlier than constraint is considered a "hard" constraint. In general, backward scheduling is not a best practice because it removes all float from a path of activities.

USING DATE CONSTRAINTS

As noted previously, date constraints are generally used to demonstrate an external event's effect on a schedule. However, because they prevent activities from responding dynamically to network logic, including actual progress and availability of resources, they can affect float calculations and the identification or continuity of the critical path and can mask progress or delays in the schedule. Date constraints should be minimized because they restrict the movement of activities and can cause false dates in a schedule. They can also imply a false level of criticality because of their effects on float. Moreover, constraints affect the analysis of risk in the schedule. Hard constraints can sometimes be impossible to meet, given the network's characteristics, and can thereby result in schedules that are logically impossible to carry out.

SNET constraints are valuable when an activity cannot start any earlier than a fixed date and has no other logical dependencies. They are often used to represent the availability of cash flow or reliance on some external product. For example, a production line may not be available until an outside entity finishes producing its product. In that case, a SNET constraint would legitimately prevent scheduled activities from unrealistically starting before they should.

SNET constraints are also used to signal the receipt of some item, such as a hardware subcomponent or a government-furnished test article. However, often these conditions of supply by an outside vendor or contractor are better represented as actual activities in the schedule. For example, the receipt of a subcontractor's hardware component is often modeled as a milestone with a SNET constraint. It may be more appropriate to model this as one activity or a sequence of activities representing the whole procurement process, starting with ordering the product and ending with its receipt.

These types of representative activities—referred to as "reference" activities or "schedule visibility" activities—are not part of the project scope and have no assigned resources, yet they can potentially affect project activities. For example, the time to produce the product should be represented by a fabrication activity. By modeling vendor or contractor production as an activity, the program office can track the high-level progress of the contractor and apply risk to the external production activity. Representing supplier, government, or subcontractor deliverables as date-constrained milestones rather than activities may understate the risk in the procurement process.

SNET constraints are also often used to delay activities in response to available resources such as labor or funding. However, this model should not be used for several reasons. First, it prevents the constrained activity from dynamically taking advantage of possible time savings being produced by predecessor activities. Second, logical dependencies should not be used to allocate resources because, typically, resource-constrained activities that are resolved with date constraints are forced to occur sequentially. This may temporarily solve a specific resource overallocation, but the sequential logic will remain even if additional resources are assigned to the activities.

Finally, perhaps the main disadvantage of applying SNET constraints to represent the availability of resources is that it requires constant manual upkeep of the schedule. Updating constraint dates on associated activities manually may be manageable in a relatively small schedule with few resources. However, large schedules with hundreds of SNET constraints representing tens or hundreds of resources will quickly become unmanageable, and the likelihood of errors will increase. If decision makers are not aware that an unnecessary SNET constraint on a low-level detail activity is preventing the activity from starting earlier, additional opportunities for time savings will be lost.

As described in Best Practice 3, resources should be assigned to activities so that their availability can drive the dates of planned activities according to resource calendars. For example, suspensions of work because of weather events are more appropriately indicated by a nonworking period in the working calendar than by date constraints. Updating resource calendars is easier to manage than manually updating hundreds of individual constraints (see case study 6).

Case Study 6: Managing Resources with Constraints, from *DOD Business Transformation*, GAO-11-53

GAO's analysis of the Army's Global Combat Support System found that dependencies within the schedule were generally sound, but 60 percent of the activities (or 1,360) had Start No Earlier Than (SNET) constraints. SNET constraints are considered "soft" date constraints because they allow an activity to slip into the future if their predecessor activity is delayed, but the activity cannot begin earlier than its constraint date.

Program officials stated that SNET constraints were used to manually allocate resources and coordinate data tests, which relied on coordination with outside partners. Officials further stated that individual control account managers monitor these constraints. GAO found that 87 percent of the constraints were actively affecting the start date of their activities. That is, without the constraint, it might have been possible to start the activity sooner.

Constraining over half of all activities to start on or after specific dates defeats the purpose of a dynamic scheduling tool and greatly reduces the program's ability to take advantage of possible time savings.

GAO, *DOD Business Transformation: Improved Management Oversight of Business System Modernization Efforts Needed*, GAO-11-53 (Washington, D.C.: October 7, 2010).

More information on resources and calendars is given in Best Practice 3 and Best Practice 4.

FNET constraints are used to prevent activities from ending too early. These constraints should be questioned because it is usually not clear why a manager would not want to end a task as soon as possible. Like SNET constraints, FNET constraints may be used in situations to force the allocation of resources. For example, preventing an activity from ending too soon can prevent a dedicated resource from going idle if a succeeding activity cannot employ that resource.

But also like SNET constraints, FNET constraints prevent a planned activity from taking advantage of any time savings that may be created by completing predecessor activities early. Because the FNET constraint sets the early finish date, the activity's early start becomes dependent on its early finish and duration. In addition, total float calculations can become obscured because of the artificial early finish date. That is, there may be time to start the activity earlier than originally planned if its duration estimate grows, but this may not be obvious given its available float.

Finally, SNET and FNET date constraints that are not actively constraining an activity's start or finish date should be removed from the schedule. In these cases, the constraints are meaningless.

The hard MSON and MFON constraints prevent activities from moving either forward or back in the plan in response to the status of predecessor activities. This includes preventing the constrained activities from taking advantage of time savings from predecessor activities. In other words, even if the constrained activity could start earlier, it will not do so according to network logic because the early and late dates are equal.

Placing a hard constraint on an activity fixes the date and immediately causes the activity to become critical. It is therefore possible to use hard constraints as a temporary working tool during schedule development to calculate total available float up to key milestones. The temporary use of hard constraints is also valuable for assessing the realism of available resources to achieve the planned activity date. For example, a hard constraint placed on an intermediate delivery milestone may show the need for an immediate and unrealistic peak of resources, shortening the predecessor durations because it is forcing the milestone to be achieved on an unrealistic date.

Hard constraints are useful for calculating the amount of float available in the schedule and, therefore, the realism of the required program finish date and available resources during schedule development. However, they may be abused if they force activities to specific dates that are determined off-line without much regard for the realism of the assumptions necessary to achieve them. It is important to note that just because an activity is constrained in a schedule, the activity is not necessarily constrained in reality. A customer-mandated date, including contractual obligations, does not constitute a legitimate reason to constrain an activity. A schedule is intended to be a dynamic, pro-active planning and risk mitigation tool that models the project and can be used to track progress toward important program milestones. Schedules with constrained dates can portray an artificial or unrealistic view of the program and begin to look more like calendars than schedules. Such unrealistic views are especially dangerous when they are translated to higher-level summary schedules for decision makers' use.

Senior management may not be aware that key milestone dates in the summary schedules are artificially fixed and behind schedule in working-level detailed schedules. Working versions of schedules may include hard constraints to assess available float and available resources, but the baseline schedule and official status updates should not contain hard constraints. If a hard constraint cannot be avoided, it must be used judiciously and must be fully justified by referring to some controlling event outside the schedule in the schedule's basis document.

In summary,

- SNET and FNET constraints delay activity starts and finishes even if predecessor durations allow them to occur earlier. These constraints allow activities to slip if their predecessors cause them to slip—in this case, they become meaningless and should be deleted. However, their use must be justified because, in general, program management's wanting to not start or finish an activity as soon as possible may be questionable.
- Because SNLT and FNLT constraints prevent activities from slipping, their use is discouraged. They should never appear in the schedule baseline. If they are not properly justified in working schedules, they must be immediately questioned.
- Because MSON and MFON constraints prevent activities not only from slipping but also from accelerating, their use is discouraged. They should never appear in the schedule baseline. If not properly justified in working schedules, they must be immediately questioned.

Many activities with date constraints are typically substitutes for logic and can signify that the schedule was not well planned and may not be feasible. In addition, constraints can cause misleading risk analysis results because they override network logic.

LAGS AND LEADS

A lag in a schedule denotes the passage of time between two activities. Lags simply delay a successor activity—no effort or resources are associated with this passage of time. A common example of the use of a lag is the passage of time to allow concrete to cure, as in figure 17. The forms on the basement walls cannot be stripped until the concrete cures. While other work may occur in parallel in other parts of the house, this particular sequence of activities is delayed so the concrete can cure. Without the lag, work associated with stripping the forms may start as soon as the basement walls are poured. The lag delays the start of the successor activity to allow time for the concrete to cure.

Figure 17: A Lag

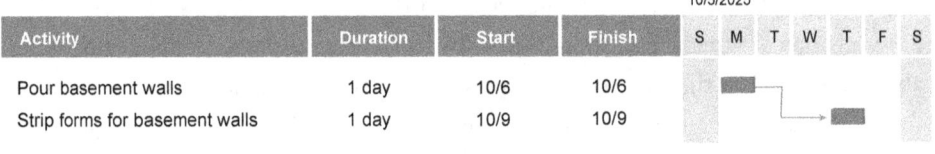

Activity	Duration	Start	Finish	S	M	T	W	T	F	S
Pour basement walls	1 day	10/6	10/6							
Strip forms for basement walls	1 day	10/9	10/9							

Source: GAO. | GAO-16-89G

A negative lag, known as a lead, is used to accelerate a successor activity. In figure 18, installing the exterior siding and brick finishes is accelerated to begin 1 day before the installation of the exterior doors and windows is finished. Without the lead, installation of the exterior siding and brick finishes would start as soon as the exterior doors and windows were installed. The lead accelerates the start of the exterior siding and brick finishes installation activity and allows both installation activities to overlap by a day. If this is possible, the project might finish ahead of schedule.

Figure 18: A Negative Lag (or Lead)

Activity	Duration	Start	Finish
Install exterior doors and windows	4 days	11/13	11/18
Install exterior siding and brick finishes	5 days	11/18	11/24

Source: GAO. | GAO-16-89G

USING LAGS

Like constraints, lags have a specific use in scheduling but may be abused. Because lags denote the passage of time, they are often misused to force successor activities to begin on specific dates. For example, if the general contractor has directed that the installation of roof trusses must start on November 6, yet its predecessor activity finishes on November 3, a 2-working-day lag would force the installation activity to occur later than necessary (figure 19).

Figure 19: Using a Lag to Force an Activity's Start Date

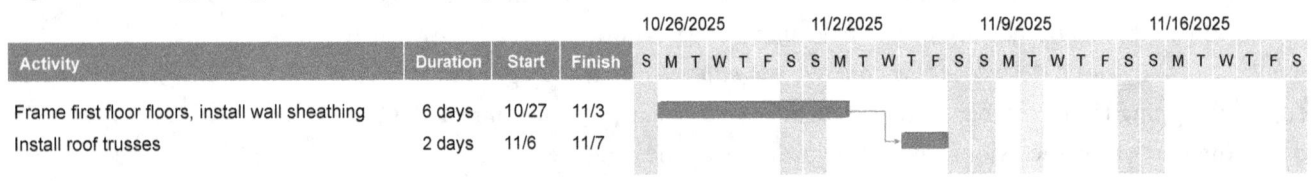

Activity	Duration	Start	Finish
Frame first floor floors, install wall sheathing	6 days	10/27	11/3
Install roof trusses	2 days	11/6	11/7

Source: GAO. | GAO-16-89G

The lag in figure 19 lessens the ability of the project schedule to dynamically respond to changes in the status of predecessor activities. A lag is static; that is, the lag will always be 2 days long unless the scheduler manually changes it. As shown in figure 20, if the predecessor activity is extended 1 day, the 2-day lag forces the mandated event to occur

1 working day later than November 6. For the event to occur on November 6, the lag must be changed to 1 day.

Figure 20: The Effect of a Lag on Successor Activities

Activity	Duration	Start	Finish
Frame first floor walls, install wall sheathing	6 days	10/27	11/4
Install roof trusses	2 days	11/7	11/10

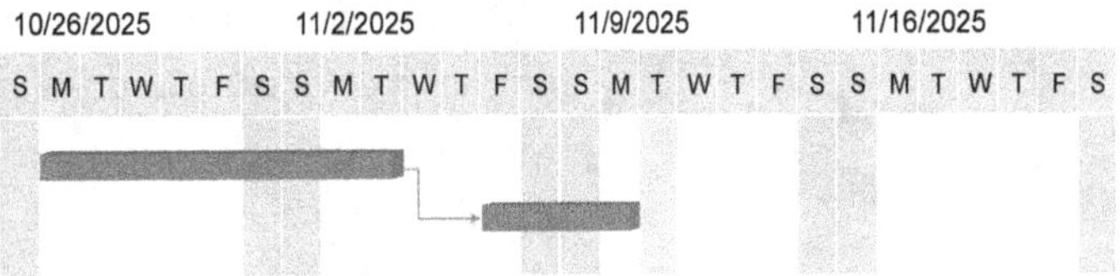

Source: GAO. | GAO-16-89G

Constantly updating manually defeats the purpose of a dynamic schedule and can make a schedule particularly prone to error when it contains many lags. If for some compelling reason outside the schedule, the installation of roof trusses cannot start before November 6, the better approach would be to put a SNET constraint there. Hence, that date will be maintained unless its predecessor is delayed beyond November 5, and then the activity will be pushed to a later date by logic.

Lags, like date constraints, must be used judiciously. They represent a real need to delay time between two activities. They must be justified by compelling reasons outside the schedule in the schedule documentation. In particular, F–S relationships with lags are generally not necessary. In these cases, every effort should be made to break activities into smaller tasks and to identify realistic predecessors and successors so that logic interface points are clearly available for needed dependency assignments. If used improperly, lags can distort float calculations in a schedule and can corrupt the calculation of the critical path. Lags are useful in summary and intermediate schedules because portions of long-term effort are likely to be unknown, or one may wish to reduce the number of activities displayed in high-level reports. Lags are usually used in places where detail is not sufficient to identify the needed interface points for making proper relationships, as in early summary-level schedules.

But lags must not be used in the place of effort in detailed schedules because lags use no resources. Likewise, lags should not represent procurement activities or other types of work performed by parties external to the schedule. Because lags are static and simply denote the passage of time, they cannot be updated with progress, are not easily monitored, and do not respond to the availability of resources. Reference or schedule visibility activities are more appropriate to represent effort that is not part of the scope yet has the potential to affect program activities.

Lags are also often used as buffers or margin between two activities for risk. However, this practice should be discouraged because the lags persist even as the actual intended margin is used up.

USING LEADS

Leads, in particular, are often unnecessary. As negative lags, leads imply the unusual measurement of negative time and exact foresight about future events. In the example in figure 18, management must assume some prescience to start installing siding and brick finishes 1 day before doors and windows are installed. Any delay in installing doors or windows within the last day may cause immediate problems for the plan. A 2-day look ahead might seem trivial, but leads of weeks or even months are unfortunately common in some schedules, questioning people's credibility when forced to see that far into the future on a risky project. In effect, a lead indicates that a future event will dictate the timing of an event in the past, which is neither logical nor possible.

The concept of a lead is better represented by a positive lag on an S–S relationship or, even more straightforward, on F–S relationships with no lags using activities of shorter duration. For example, instead of a lead, the installing doors and windows activity should be broken into smaller activities to identify the proper F–S activities. For example, the doors and windows activity might be broken into two sequential subactivities: "install exterior east and west doors and windows" and "install exterior north and south doors and windows." "Install exterior siding and brick finishes" could then be linked with a finish-to-start relationship to "install exterior east and west doors and windows," as shown in figure 21.

Figure 21: Eliminating Leads with Finish-to-Start Links

Activity	Duration	Start	Finish
Install exterior east/west doors, windows	2 days	11/13	11/14
Install exterior north/south doors, windows	2 days	11/17	11/18
Install exterior siding and brick finishes	5 days	11/17	11/21

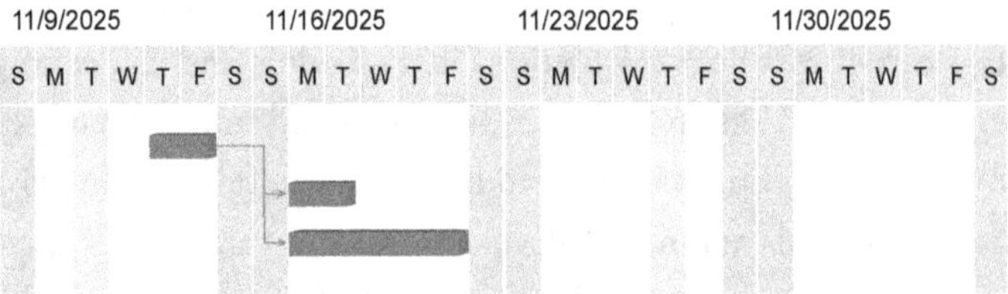

Source: GAO. | GAO-16-89G

The network now clearly shows that once the east and west exterior doors and windows have been installed, installing siding and brick finishes can begin. It also clearly identifies that siding and brick finishes installation can go on while doors and windows are being installed in the last two sections of the house. Finally, using leads improperly can cause logic failures when a lead is longer than the successor activity. An example is given in figure 22.

Figure 22: Logic Failure Associated with a Lead

Activity	Duration	Start	Finish
Finish drywall (tape and mud)	5 days	12/8	12/12
Apply drywall texture	3 days	12/9	12/11

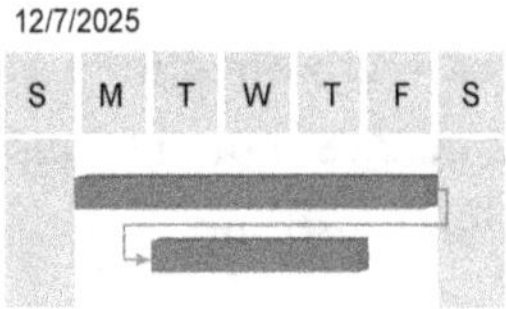

Source: GAO. | GAO-16-89G

The F–S link in figure 22 between "finish drywall" and "apply drywall texture" dictates that finishing the drywall must finish before applying the drywall texture can begin. Suppose the relationship also has a 4-day lead that states that applying the drywall texture should begin 4 days before drywall is finished. However, applying texture takes only 3 days, which means this activity will actually finish before the drywall is finished. This causes a conflict in logic between the lead and the finish-to-start relationship. Moreover, because the lead causes texture application to finish a day earlier than the finishing of drywall, an artificial day of float may be introduced into the sequence of activities.

Note that the number of activities affected by a lag or lead is different from the total number of lags or leads in the schedule. The total is useful for determining the extent to which updating the schedule will be affected by the use of lags and leads. As stated above, a major disadvantage of lags and leads is that they are static. If they are prevalent in the schedule, the update process requires significant manual effort and becomes time consuming and prone to error.

The total number of predecessor relationships with a lag or lead provides a scheduler with a sense of how cumbersome updating will be. If the schedule is not well maintained, its utility as a dynamic management tool will be reduced and activity dates may be artificially delayed or accelerated by a static lag or lead. For this reason, it is also worthwhile to understand the number of activities that will be affected by such a situation. For example, figure 23 gives an example of three concrete-pouring activities. Each activity has a 2-day lag from its predecessor to represent the time needed for curing. But because "form and pour driveway" takes a day longer than "form and pour sidewalks" and "form and pour patio," only its lag directly affects the start date of "finish grade property."

Figure 23: Enumerating Lags

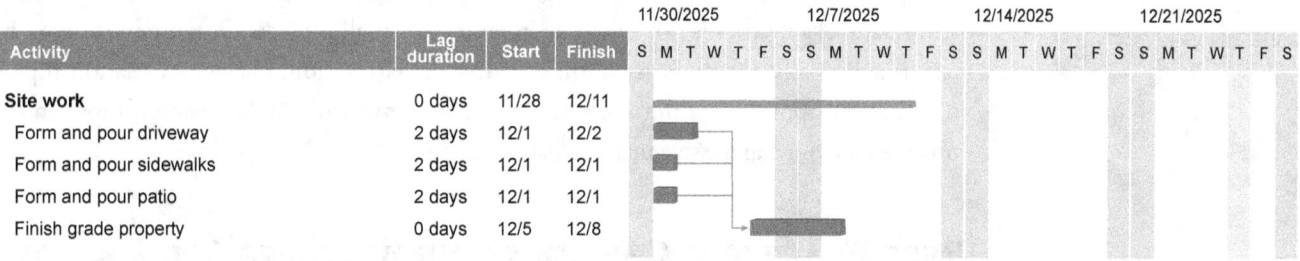

Activity	Lag duration	Start	Finish
Site work	0 days	11/28	12/11
Form and pour driveway	2 days	12/1	12/2
Form and pour sidewalks	2 days	12/1	12/1
Form and pour patio	2 days	12/1	12/1
Finish grade property	0 days	12/5	12/8

Source: GAO. | GAO-16-89G

PATH CONVERGENCE

Joining several parallel activities with a single successor activity is known as path convergence. Path convergence can be unrealistic in a plan because it implies the need to accomplish a large number of activities on time before a major event can occur as planned. The convergence of many parallel activities into a single successor—also known as a "merge point"—causes problems in managing the schedule.[14]

These points should be a key program management concern because risk at the merge point is multiplicative. That is, because each predecessor activity has a probability of finishing by a particular date, as the number of predecessor activities increases, the probability that the successor activity will start on time quickly diminishes to zero. Path convergence is the basis of "merge bias," which we discuss in detail in Best Practice 8.

Because of this risk effect, activities with a great many predecessors should be examined to see if they are needed and if alternative logic can be used to link some predecessors to other activities. Predecessor activities should also be examined for available float. If many of the predecessors leading to the merge point have large amounts of float available, then convergence may not be an immediate issue. However, if several predecessor activities are determining the date of the successor event, then the workflow plan should be reexamined for the realism of performing many activities in parallel with the available resources.

Predecessors with large amounts of total float that lead to a merge point may indicate that activities are not sequenced correctly or optimally. Often paths converge because major milestones are used to "tie off" many predecessor activities, some of which may be only marginally related to the actual milestone.

The appropriate number of converging activities varies by project. The reasonable number of parallel activities is not the same in large and small projects. Because most work is performed serially, the majority of the schedule activities should have F–S relationships, in a waterfall approach to the work. An excessive number of parallel relationships can indicate an overly aggressive or unrealistic schedule.

BEST PRACTICES CHECKLIST: SEQUENCING ALL ACTIVITIES

- The schedule contains complete network logic between all activities so that it can correctly forecast the start and end dates of activities within the plan.

- The majority of relationships within the detailed schedule are finish-to-start.

---

[14]Parallel lines do not converge in mathematics. However, in scheduling parlance, paths and activities are parallel if they occur in the same work periods. The paths converge when their activities have the same successor.

- Except for the start and finish milestones, every activity within the schedule has at least one predecessor and at least one successor.

- Any activity that is missing predecessor or successor logic—besides the start and finish milestones—is clearly justified in the schedule documentation.

- The schedule contains no dangling logic. That is,

   ○ Each activity (except the start milestone) has an F–S or S–S predecessor that drives its start date.
   ○ Each activity (except the finish milestone and deliverables that leave the program without subsequent effect on the program) has an F–S or F–F successor that it drives.

- The schedule does not contain start-to-finish logic relationships.

- Summary activities do not have logic relationships because the logic is specified for activities that are at the lowest level of detail in the schedule.

- Instead of SNET constraints, conditions of supply by an outside vendor or contractor are represented as actual activities in the schedule.

- Date constraints are thoroughly justified in the schedule documentation. Unavoidable hard constraints are used judiciously and are fully justified in reference to some controlling event outside the schedule.

- Lags are used in the schedule only to denote the passage of time between two activities.

- Every effort is made not to use lags and leads but to break activities into smaller tasks to identify realistic predecessors and successors so that logic interfaces are clearly available for needed dependency assignments.

- Lags and leads in a schedule are used judiciously and are justified by compelling reasons outside the schedule in its documentation.

- The schedule is assessed for path convergence. That is, activities with many predecessors have been examined to see whether they are needed and whether alternative logic can be used to link some predecessors to other activities.

BEST PRACTICE 3

Assigning Resources to All Activities

Best Practice 3: The schedule should reflect the resources (labor, materials, travel, facilities, equipment, and the like) needed to do the work, whether they will be available when needed, and any funding or time constraints.

A resource is anything required to perform work. Because resource requirements directly relate to an activity's duration, assigning resources to activities ensures that the duration of activities using them will be realistic and rational. Because labor, material, equipment, burdened rates, and funding requirements are examined to determine the feasibility of the schedule, resources provide a benchmark of the program's total and reporting-period costs.

The process of assigning labor, materials, equipment, and other resources to specific activities within the schedule is known as loading resources. A resource-loaded schedule therefore implies that all required labor and significant materials, equipment, and other costs are assigned to appropriate activities. The schedule should realistically reflect the resources that are needed to do the work and—compared to total available resources—should determine whether all required resources will be available when they are needed.

Loading resources is the first step in allocating the expected available resources to a schedule at the time planned for performing an activity. Representing all resources in an IMS may be difficult for complex programs. But as noted in Best Practice 1, the more complex a program is, the more complex the IMS may become. An analyst must examine resources by time period to determine whether they are adequate. To represent the total and reporting-period costs of the project, all resources (including subcontracts and LOE management resources) must be included. A schedule that has not been reviewed for resource issues is not reliable.

RESOURCES, EFFORT, AND DURATION

The amount of available resources, whether labor or nonlabor, affects estimates of work and its duration, as well as resources available for subsequent activities. Labor is, of course, a human resource; nonlabor resources can be subcontracts, consumable material, machines, and other purchased equipment. Labor and equipment are measured in units of time such as hours and days; material resources are measured in cubic feet, tons, pallets, or the like. Resources can depend on time—generally labor but also equipment or space rented per period—meaning that they increase in cost as time runs longer. Alternatively, they can be time-independent in the sense that they cost the same regardless of time—for example, material costs. Some activities use both labor and nonlabor resources.

Labor Resources

In general, the amount of work, in person-days, required to complete an activity is equal to the duration of the activity multiplied by the number of labor resource units assigned. Stated another way, the duration of an activity is directly related to how much work is necessary to complete the activity divided by the number of people available to perform the work. However, this does not necessarily mean that doubling available resources will halve the activity's duration.

If the amount of work is known, along with an estimate of the number of people available to perform that work, then its duration can be estimated, along with efficiency levels, risk, and other external factors. That is, when the amount of work is fixed, the number of labor resources directly affects its duration. For example, if an activity is estimated to require 16 hours of work (2 person-days) and only one full-time equivalent (FTE) employee is available to perform the activity within an 8-hour day, then the duration of the activity will be 2 days. If two FTEs are available to perform the activity, then the duration will be 1 day; if one FTE is available for only 50 percent of the time, the duration will be 4 days.

Productivity or efficiency factors can be applied to specific resources to account for standard output rates, personal experience, or historical productivity. For example, specific persons in a resource group may have more experience performing an activity than other persons in the same group. Productivity varies by the type of work—for example, trees planted per day or drawings produced per week. Workflow also affects productivity: a steady flow of work tends to increase efficiency, whereas a discontinuous flow can introduce inefficiencies. Finally, complex activities may actually require additional duration as more people are assigned to account for greater communication and coordination requirements.

The duration of other types of activities known as fixed-duration activities is not affected by the number of people assigned to perform the work. For example, the number of days required for testing a satellite in a vacuum chamber will be the same regardless of how many engineers are assigned to monitor the testing. Likewise, the duration of a management offsite meeting does not depend on the number of people who attend. In the case of fixed-duration activities, labor resource information is nonetheless vital because the number of people directly affects the work required for the activity and, therefore, the cost. For example, five engineers assigned to a 2-day fixed-duration software coding activity will incur 10 person-days of work at whatever labor rate each engineer earns.

Nonlabor Resources

Significant material and equipment resources should also be specified within a schedule. Material resources are consumables and other supplies that are used to complete a project. Equipment resources may be items such as machines that are installed during the project and become part of the completed project at turnover. Other equipment resources may facilitate a project's execution but are neither consumed nor turned over in the final product delivery—for example, a rented crane.

Like labor resources, nonlabor resources can be fixed or variable. Fixed nonlabor resources do not vary with an activity's duration; that is, the same amount of resource is consumed regardless of the activity's duration. For example, 100 square feet of wood flooring might be needed for a floor, regardless of whether the floor construction takes 2 or 3 days to complete. Variable nonlabor resources may include equipment used during the project such as cranes or testing machines that are not consumed but provide services that vary with the duration of an activity. For example, the longer a testing activity runs, the longer the test equipment is in use.

ROLLING WAVE PLANNING

As discussed in Best Practice 1, a comprehensive IMS should reflect all the activities of a program and should recognize that uncertainties and unknown factors in schedule estimates can stem from, among other things, limited data. A schedule incorporates different levels of detail depending on the information available at any point in time. Near-term effort will be planned in greater detail than long-term effort.

Detailed activities within a low-level schedule represent tasks that are typically 4 to 8 weeks long.[15] They reflect near-term, well-defined effort, typically within 6 months to a year of the current date. But it is often difficult to forecast detailed work clearly beyond 9 to 12 months. In general, the length of the near-term detail planning period should be decided by program management. It depends on the project's size, phase, scope, risk, and complexity. For example, some schedules may be planned in detail for only 2 or 3 months.

Effort beyond the near term that is less well defined is represented within the schedule as planning packages. Planning packages summarizing work in the distant future can be used as long as they are defined and estimated as well as possible. Planning packages are planned at higher levels such that a single activity may represent several months of effort, generic work to be accomplished by a trade or resource group, or even a future contract or phase.

As time passes and future elements of the program become better defined, planning packages are broken into detailed work packages. This incremental conversion of work

---

[15] Durations of detail activities are discussed in Best Practice 4.

from planning packages to work packages is commonly known as "rolling wave" planning. Rolling wave planning continues for the life of the program until all work has been planned in detail.[16] A best practice is to plan the rolling wave to a design review, test, or other major milestone rather than to an arbitrary period such as 6 months.

Moreover, detail should be included in the schedule whenever possible. That is, if portions of far-term effort are well defined, they should be included in the IMS as soon as possible. However, care should be taken not to detail ill-defined far-term effort so soon as to require constant revision as time progresses. More detail does not necessarily mean greater accuracy, and pursuing too much detail too early may be detrimental to the schedule's quality.

While planning packages represent far-term effort that has not yet been planned in detail, each planning package must still be traceable to WBS elements within the IMS. Moreover, planning packages should be logically linked within the schedule to create a complete picture of the program from start to finish and to allow the monitoring of a program's critical path. Planning packages that are on or near the critical path or that carry significant risk should be broken into smaller activities to better understand workflow. As durations and resource assignments are refined over time, so too is the detailed sequence of activities.

Appendix III provides more information on work packages, rolling wave planning, and earned value management.

LOADING ACTIVITIES WITH RESOURCES

Including resources in a schedule helps management compute total labor and equipment hours, calculate total project and per-period cost, resolve resource conflicts, and establish the reasonableness of the plan. A schedule without resources implies an unlimited number and availability of resources. Resource information can be stored within the schedule files or it can be stored externally in separate software, but a best practice is to store resources in the schedule itself.

Fully loading the schedule with resources, including materials, equipment, direct labor, travel, facilities, equipment, and level-of-effort activities, provides the basis for the performance measurement baseline (PMB), which can be used to monitor the project using earned value management (EVM).[17] When a schedule is fully resource loaded, budgets for direct labor, travel, facilities, equipment, material, and the like are assigned to both

---

[16]Rolling wave planning with portions of effort that align to significant program increments, blocks, or updates is sometimes referred to as "block planning." The distinction is that rolling wave planning is performed with a certain periodicity while block planning is performed to specific stages in the project. In either case, details are added to the schedule incrementally as the project progresses.

[17]The performance measurement baseline is the time-phased budget plan for accomplishing work. For more information on the relationship between cost, schedule, and EVM, see appendix III.

work and planning packages so that total costs to complete the program are identified at the outset.

How detailed resource assignments and duration estimates are varies by how detailed activities are in the schedule. Work scheduled for the near term should account for specific resource availability and productivity. The level of detail for resource loading should be based on the information available to management as well as the purpose of the schedule. Detail schedules used for day-to-day management should include detail resource information, while summary or intermediate level schedules with higher-level resource information may be more appropriate for analyses such as an integrated cost-schedule risk analysis.

Because milestones have no duration, they should never be assigned resources. Additionally, for the scheduling software packages that include them as an option, summary activities should not be assigned resources. Summary activity durations depend on the activities contained within them.

Activity owners responsible for managing the day-to-day effort and the most experienced team members who will be performing the work are the best source of resource estimates. Activity owners must be able to explain the logic behind their resource estimates; if there is no justification for allocating and assigning resources, the schedule will convey a false accuracy. Estimated resources within the schedule should also reconcile with the cost estimate. The assumptions for resources and related activity cost should be the same as those that are used in estimating activity duration. The basis of estimate (BOE) is the connection between cost and time and should be kept up to date as assumptions change. If durations, resources, or productivity rates change, the cost is also likely to change, and they need to be coordinated. Both the schedule and the cost estimate should be thoroughly documented to include underlying resource assumptions for the entire estimated scope of work.

In terms of labor resources, a schedule can be loaded at different levels of detail, each level having its advantages and disadvantages. In some cases, a program may rely on a centralized resource pool, previously approved by management, that must be used as the detailed work schedules are populated. Assigning labor resources to activities at the name level ("Bob Smith Jr.") is specific and allows management to track the effort of individual people and to make sure they are not overallocated. This is particularly useful for tracking the assignments and responsibilities of subject matter experts whose absence or overallocation could delay an activity. However, assigning individual names to activities requires constant updating and shuffling of assignments for such events as vacations and sick leave and it reduces flexibility in resource leveling. Individual name assignments may also require specific labor rates, although planners may simply enter the average rate for each employee to obviate the need for this.

Assigning resources at the organizational level such as "agency test department" or "production facility staff" is quicker than assigning individual names to activities. However, organizations include multiple skill sets, which may complicate allocating and assigning specific skill sets to activities. Assigning resources by skill set ("software engineer level 3") may be more time consuming than organization-level assignments but is a compromise option for resource loading to avoid both too much and too little detail. Skill sets obscure specific labor rates, parse individuals' skills, and allow for management's rearrangement of individual personnel within skill sets.

Resource information can be stored within the schedule files or externally in separate software such as in systems devoted to financial management or manufacturing resource planning.[18] If resource information is stored and maintained outside the schedule, how information is integrated between the schedule and the resource management software must be clear. Specifically, managing resources outside the schedule requires a procedure in which resource assignments are fed back into the schedule to reflect the separate resolution of resource issues. A best practice is to store the resource information in the schedule. Unless resources are stored in the schedule itself, the effects of resource availability must be manually updated, increasing the likelihood of error. In addition, managers and auditors can more easily identify and verify critical resources if resources are stored in the schedule file. Finally, if resources are not stored in the schedule, resource leveling will be difficult.

Case study 7 provides an example of a fully resource-loaded schedule and its alignment to the program budget baseline.

Case Study 7: Assigning Resources to Activities, from *Nuclear Nonproliferation*, GAO-10-378

GAO assessed the extent to which the project schedule for the Department of Energy's Savannah River Waste Solidification Building used best practices. In addition to fully reflecting both government and contractor activities, the sequencing of 95 percent of the remaining 2,066 activities, and the establishment of a credible critical path driven by the logic of the schedule, the project's schedule included assigned labor and material resources that accounted for 98 percent of the project's $344 million cost baseline. By fully meeting the best practice of assigning resources to activities, the program office was better able to assess resource availability during work periods. In addition, assigning resources to all activities meant that management acknowledged that resources were limited, allowing it to identify and rectify overallocated resources.

GAO, *Nuclear Nonproliferation: DOE Needs to Address Uncertainties with and Strengthen Independent Safety Oversight of Its Plutonium Disposition Program*, GAO-10-378 (Washington, D.C.: March 26, 2010).

---

[18]Manufacturing resource planning (MRP) is a software system designed to improve resource planning by using an automated information management system that provides planning, scheduling, capacity, and other information.

After the schedule is resource loaded, total resources in the schedule should be cross-checked with the program budget and contractual cost constraints. Crosschecking can be performed at the summary level by rolling up lower levels of the schedule to key high-level milestones.

The cumulative resources and budget can be visualized graphically as a series of peaks and troughs of resource and funding availability. Resource peaks should be examined for feasibility regarding the available budget, availability of resources, and timeliness. For example, scheduled activities may call for 70 software engineers, but actual facility or administrative constraints may prevent the staffing of more than 40 software engineers simultaneously. The benefits of using a schedule to reallocate resources can be seen in the example of house construction in figure 24.

Figure 24: A Profile of Expected Construction Labor Costs by Month

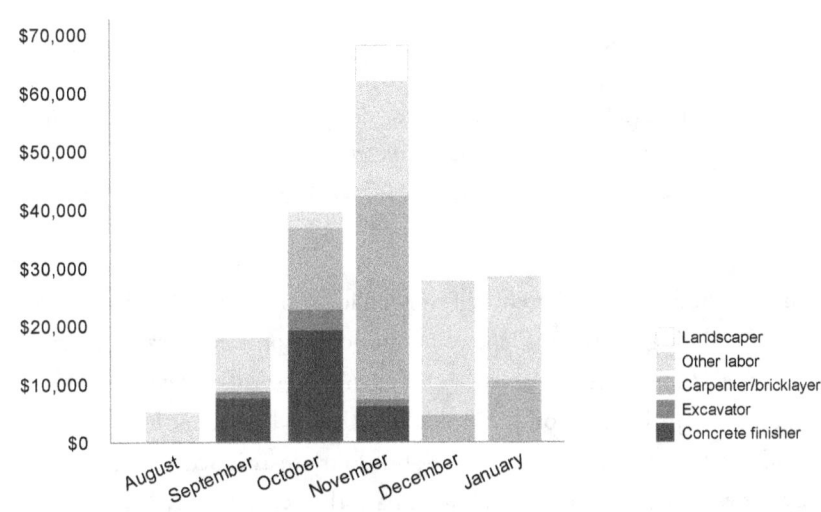

Source: GAO. | GAO-16-89G

The expected labor costs peak severely in November, corresponding with a surge in carpentry and bricklaying work, as well as final grading, landscaping, and pouring concrete for the patio, driveway, and sidewalks once the exterior house work is complete. The labor budgets then drop substantially for December and January, creating potential cash flow issues for the owners or the general contractor. However, final grading and landscaping need be completed only by the time of the owner's walkthrough, not scheduled until late January. As a consequence, those activities have over 30 days of total float available; that is, the final landscaping and concrete pouring activities can be delayed until December without delaying the owner's acceptance. This shifts nearly $14,000 in labor costs from November to December, easing the budget concerns in November and allowing work and cash flow to ramp down smoothly into January (see figure 25).

Figure 25: A Smoothed Profile of Expected Construction Labor Costs by Month

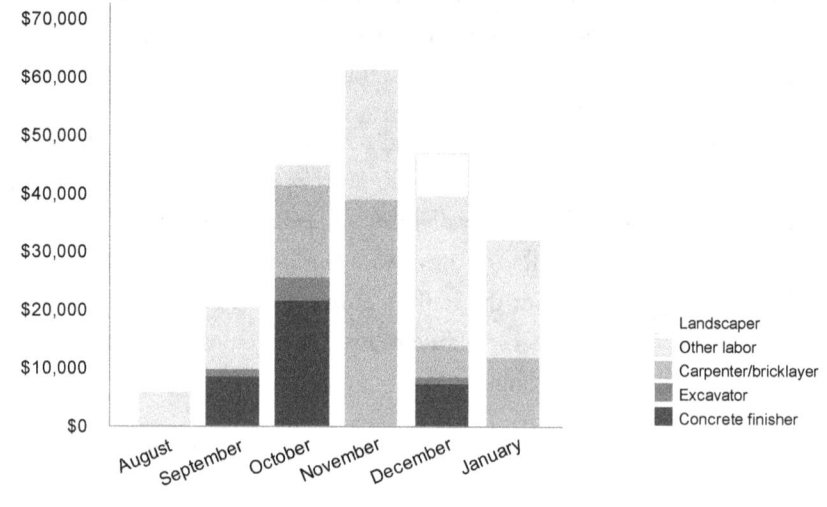

Source: GAO. | GAO-16-89G

Peaks in the resource profile often occur before major decision points in the program. If the cumulative overlay of resources against major milestones shows resource peaks just beyond major milestone points—rather than just before—then resources may have to be reallocated.

Resource assignments should also be carefully reviewed for realism in light of assumptions made by the scheduling software. Depending on how resources are assigned and durations are estimated, the scheduling software may ignore resource overallocations by allowing assignments to continue beyond resource availability. For example, if only three carpenters are available to perform cabinetry work but four are assigned to the activity, the software may allow four carpenters and simply mark the activity as overallocated. In reality, a fourth carpenter is not available and the duration of the activity would have to be extended.

While resource loading the entire schedule may be a difficult exercise, it encourages management to assess the amount of resources available and encourages a discussion of difficult questions early in program planning. If the resource-loaded schedule alerts decision makers that the available resources will not suffice to execute the work on time as planned, management can begin negotiating for additional resources early in the program. Finally, linking available resources to activity durations may expose infeasible durations to scrutiny or show opportunities to reduce durations with the application of more resources. See Best Practice 4 for more information on establishing durations for activities. Case study 8 illustrates how optimistic resource assumptions can delay a program plan.

Case Study 8: Assuming Unlimited Availability of Resources, from *Arizona Border Surveillance Technology Plan*, GAO-14-368

Our analysis of the Integrated Fixed Tower (IFT), Remote Video Surveillance System (RVSS), and Mobile Surveillance Capability (MSC) project schedules related to the Arizona Border Surveillance Technology Plan found that two minimally met and one did not meet the best practice of assigning resources to activities. In particular, our analysis concluded that the MSC schedule did not meet the best practice because it was not resource loaded at any level. Program officials stated that they did not estimate resources for the MSC schedule because they developed the schedule under the assumption that the activities were to be fully resourced to do the work and that personnel would be dedicated and not shared between the other programs in the Plan.

A schedule without resource assignments implies an unlimited number of resources and their unlimited availability. If allocating and assigning resources are not justified, the schedule will convey false accuracy. This is particularly important where resources are in in multiple projects sharing critical resources. In fact, as the U.S. Customs and Border Protection's Office of Technology Innovation and Acquisition (OTIA) initiated and continued work on the Plan's programs, it shared resources such as personnel among the programs, in part delaying them. For example, with regard to the IFT program, OTIA officials stated that sharing a contracting officer with another program was necessary. Further, OTIA officials told us that because of resource constraints associated with initiating the Plan, the development of two acquisition documents—an acquisition program baseline and life-cycle cost estimate—for the MSC program was deferred because the IFT and RVSS programs were given higher priorities. In addition, for the IFT and RVSS programs, officials delayed planning and deployment activities because of resource constraints and the lack of dedicated contracting officers to plan and execute the programs' source selection and environmental activities.

GAO, *Arizona Border Surveillance Technology Plan: Additional Actions Needed to Strengthen Management and Assess Effectiveness*, GAO-14-368 (Washington, D.C.: March 3, 2014).

RESOURCE LEVELING

Resource leveling adjusts the scheduled start of activities or the work assignments of resources to account for their availability. Primarily the organization that has control of the resources uses it to smooth spikes and troughs in resource demands created by the sequencing of activities in the schedule network. Frequent peaks and troughs in resource requirements indicate their inefficient use inasmuch as frequent mobilization and de-mobilization of resources is disruptive. For example, a schedule that requires 30 software engineers one week, none the next week, and 25 the next is disruptive and inefficient. A more manageable resource plan should be sought.

Leveling can be as simple as reassigning work from overallocated resources to under-allocated resources or delaying the start date of activities until the required resources are available. Leveling may also develop into a complex trade-off between the required duration of the plan and the availability of myriad resources. Resources can be leveled automatically with scheduling software or manually by managers and planners or both. Whatever the method, the goal of resource leveling is to finish the project on time or early, if possible, with the resources realistically expected to be available throughout the entire plan. Leveling resources allows management to identify "critical resources"—that is, resources that will delay the project finish date if they are not available for specific activities. Resource leveling is ideally integrated into scheduling and updating to ensure the best possible trade-offs between resources and time and to ensure that the schedule remains reliable throughout its life.

Typically, activities delayed by resource leveling have the greatest free float available and the fewest assigned resources.[19] Leveling resources by shifting activities with free float minimizes the effect of the delayed activities on the project as a whole and minimizes the number of resources that must be reassigned work. Given the amount of float available in the schedule and the original assignment of resources, resource leveling can have little to no effect on a schedule, or it can severely delay the forecasted end date of the project. Resources should be leveled only to reallocate resources to reduce spikes and troughs by absorbing available float.

In other words, resource leveling should not delay a completion date if possible. If the outcome of resource leveling is a delayed completion date, then resources are an inevitable constraint, and an alternative approach to the project plan is needed. If resources are assigned to activities realistically according to the sequence of planned activities and no resources are overallocated, then the resource-leveled scheduled will be the same as the original CPM schedule.

An example of resource leveling is given in figures 26 and 27. Figure 26 shows a selection of the sequence of events for selecting subcontractors for house construction, as scheduled according only to predecessor-successor logic. Once the owners select a general contractor, they can confer with the general contractor to select the subcontractors: surveyor, excavator, concrete supplier, plumber, electrician, and the like. Once an individual subcontractor is selected, the owners and general contractor can select and approve the necessary material. For instance, after selecting a plumber, the owners and general contractor can select the plumbing fixtures. These activities are planned to occur while the owners wait for the general construction permit to be granted, which is expected to take 20 days. From a network perspective, all subcontractors can be selected at the same time once the owner selects the general contractor and, once each subcontractor is selected, the respective materials can be selected. This sequence of activities is portrayed in figure 26.

---

[19]Free float is the time in which an activity can be delayed without affecting any successor (see Best Practice 7).

Figure 26: Resource Overallocation in a Correctly Sequenced Network

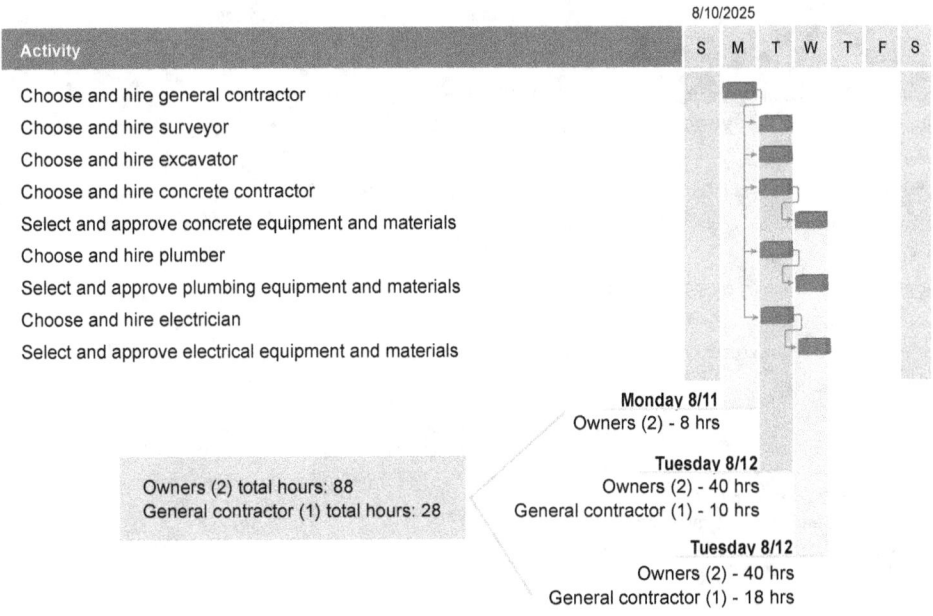

Monday 8/11
Owners (2) - 8 hrs

Tuesday 8/12
Owners (2) - 40 hrs
General contractor (1) - 10 hrs

Tuesday 8/12
Owners (2) - 40 hrs
General contractor (1) - 18 hrs

Owners (2) total hours: 88
General contractor (1) total hours: 28

However, once resources are assigned to activities, it quickly becomes apparent that the plan is not feasible. Figure 26 gives the hours necessary for two owners and one general contractor to complete the selections for each day. The resource leveling of these activities is straightforward, because the owners and the general contractor have time available to select subcontractors while they wait for approval of the general construction permit. In fact, the "subcontractors selected and equipment and materials approved" milestone has over 20 days of free float because the selection activities are occurring parallel to the wait time allotted for construction permit approval. This allows management to spread the selection of subcontractors over the next several weeks without affecting activities succeeding the subcontractor selection milestone. Figure 27 shows the results of manual resource leveling for these activities.

Figure 27: Resource Leveling in a Correctly Sequenced Network

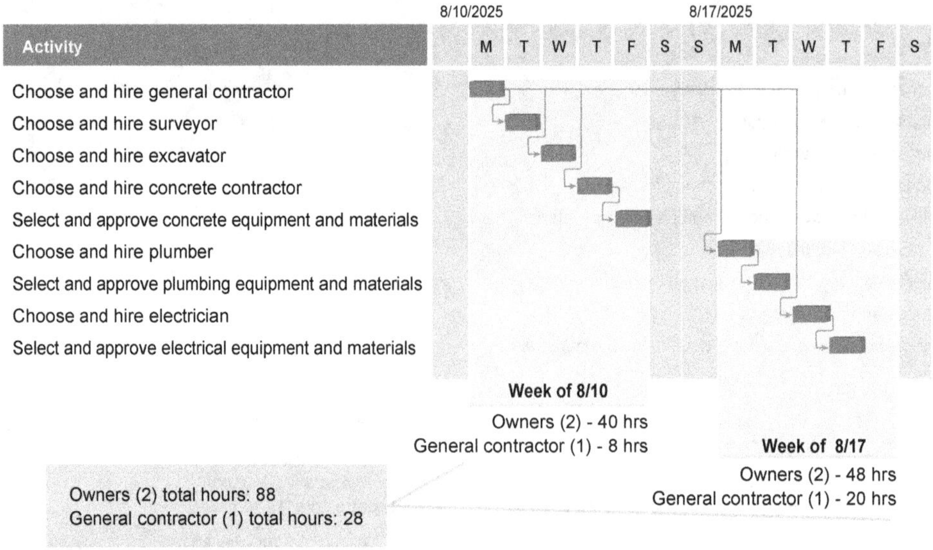

Source: GAO. | GAO-16-89G

The activities are now spread over 2 weeks, leaving the owners and general contractor busy but not overallocated. This effort profile also frees the general contractor to work with other clients. Finally, the subcontractor selection activities are leveled within the available free float, which does not affect either activities beyond the subcontractor selection milestone or the original critical path.

Because changes to the schedule from limited resource availability can alter float and critical path calculations, it is important that changes to resolve the resource conflicts be thoroughly documented and that everyone understand them.[20] Planners and management should always carefully examine output from automatic resource-leveling routines and it should be tempered or adjusted where necessary. Automatic leveling may prove inefficient, as when it delays activities when resources are only partially available and, thus, prevents activities while the project awaits the full complement of resources.

It is important to note that decisions made from incorrect data assumptions will themselves be incorrect. This is especially true if a CPM schedule is being overridden by resource-leveling decisions based on summary-level or incorrect resource assumptions. Resources should be leveled only in detailed schedules that include detailed resource estimates supported by historical data and sound estimating methodologies. Without specific resource assignments, effects and costs cannot be accurately estimated and tracked. If a schedule is resource leveled, it is important to check the date of the last

---

[20]Logic specifically employed to solve resource leveling issues once the plan is baselined is known as "preferential logic" or "soft logic." Preferential logic dictates a desired sequence of activities that is not entirely necessary. That is, the logic is not related to the work itself but reflects management's plan for realistically executing the work. Conversely, "engineering logic," or "hard logic," dictates an order of activities that must occur regardless of preference. For example, concrete curing must always occur after pouring concrete. Using preferential logic to address resource leveling may be subject to the program's schedule change control.

leveling against the date of the last changes to resources and resource availability. This will ensure that actual resource assignments have been aligned with the proposed leveling in the model.

Resource leveling should never be applied to summary schedules or when resources are specified at such a summary level that the concept of availability cannot be applied. Incorrect resource assumptions (usually in the form of unwarranted optimism) lend unreasonable credence to a resource-leveled schedule, and the resulting leveled schedule will convey a false sense of precision and confidence to senior decision makers.

BEST PRACTICES CHECKLIST: ASSIGNING RESOURCES TO ALL ACTIVITIES

- The amount of available resources, whether labor or nonlabor, affects estimates of work and duration, as well as the availability of resources for subsequent activities.

- The schedule should realistically reflect the resources that are needed to do the work and—compared to total available resources—should determine whether all required resources will be available when they are needed.

- Resources are either labor or nonlabor, where labor is tracked in hours or FTEs and nonlabor can refer to subcontracts, consumable material, machines, and other purchased equipment. Resources are identified as fixed or variable.

- Significant material and equipment resources are captured within the schedule along with other equipment resources that facilitate the project.

- Budgets for direct labor, travel, facilities, equipment, material, and the like are assigned to both work and planning packages so that total costs to complete the program are identified at the outset.

- Summary activities and milestones are not assigned resources.

- If EVM is used to monitor the program, the fully loaded schedule, including materials, equipment, direct labor, travel, and LOE activities, is the basis for the PMB.

- Activity owners are able to explain the logic behind their resource estimates.

- The same assumptions that formed resource estimates for the cost estimate are applied to the estimated resources loaded into the schedule and are documented in the BOE. Underlying resource assumptions for the entire estimated scope of work are documented in the schedule basis document in appropriate detail.

- Resource information is stored in the schedule in the form of assignments. If resource management is performed outside the schedule, a documented process feeds resource assignments back into the schedule so that it reflects the resolution of resource issues conducted separately.

- Once the schedule is resource loaded, all resources in the schedule are cross-checked with the program budget and contractual cost constraints.

- Resource peaks are examined for the feasibility of the available budget, the availability of resources, and the timeliness of the peaks. If the cumulative overlay of resources against major milestones shows resource peaks just beyond major milestone points, resources may have to be reallocated.

- Resources have been leveled—that is, the scheduled time of activities or the assignment of resources has been adjusted to account for the availability of resources.

- In general, activities that are delayed through resource leveling have the greatest free float available and the fewest resources assigned.

- If critical resources delay the entire project, changes to resolve the resource conflicts are thoroughly documented in the schedule narrative and understood by all.

- Planners and managers carefully examine and temper or adjust where necessary.

- Resources are leveled only on detail schedules that include detailed resource estimates supported by historical data and sound estimating methodologies.

BEST PRACTICE 4

Establishing the Duration of All Activities

Best Practice 4: The schedule should realistically reflect how long each activity will take. When the duration of each activity is determined, the same rationale, historical data, and assumptions used for cost estimating should be used. Durations should be reasonably short and meaningful and should allow for discrete progress measurement. Schedules that contain planning and summary planning packages as activities will normally reflect longer durations until broken into work packages or specific activities.

Duration is the estimated time required to complete an activity—the time between its start and finish. Durations are expressed in business units, such as working days, and are subject to the project calendar. For example, for a standard 40-hour 5-day work week, the duration of an activity that starts on Thursday and ends on Tuesday will be 4 working days, even though it spans 6 calendar days. If the activity is assigned to a 7-day workweek calendar, then the activity will start Thursday and end Sunday. Multiple calendars can be created to accommodate activities with different work schedules.

The definition of duration is different from the definition of work (work is also referred to as effort in some scheduling software). For example, if a painting activity is scheduled for 2 8-hour days and 2 full-time painters are assigned to the job, the duration of the painting activity is 2 days, but the effort associated with painting is 32 hours (that is, 2 8-hour employees for 2 days). Duration is directly related to the assigned resources and estimated amount of required work. Best Practice 3 discusses in detail how duration, resource units, and effort can change in relation to one another.

Durations should be as short as possible to facilitate the objective measurement of accomplished effort. As we discuss in Best Practice 1, the level of detail in the schedule should reflect the information available, the risk inherent in activities, and the intended use of the schedule. In general, estimated detail activity durations for near-term effort should be no longer than the reporting period established by the program. For example, if the reporting period for a construction project is weekly, then near-term activity durations should be one working week or less. If management requires monthly updates, then near-term activity durations should be about 22 working days or less. If activities are longer than the reporting period, activities should have at least one quantitative measurable event within the reporting period. It may be difficult for management to gauge progress on detail activity durations that are too long. Up to a point, the shorter the duration of the detail activity the more precise the measurement of accomplished effort will be. Moreover, shorter durations are needed for areas of work associated with high cost or high risk. Keeping activity durations shorter than the reporting period has additional benefits to tracking progress that we discuss in Best Practice 9.

Long durations should be broken into shorter activities if logical breaks can be identified in the work being performed. If it is not practical to divide the work into smaller activities or insert intermediate milestones, justification for long durations should be given in the schedule basis document. One rule of thumb is to break long activities into enough detail that finish-to-start logic relationships can be identified. Greater activity detail might be necessary if it helps management understand and address the implications of risk and uncertainty.

However, durations that are too short and durations that are too long should be balanced. Very short durations, such as 1 day or less, may imply that the schedule is too detailed and will require more frequent schedule duration and logic updates than necessary. Activities should be decomposed only to the point necessary to identify activity-to-activity hand-offs. Moreover, for a large number of 1- and 2-day duration activities, planners should recognize that people are rarely, if ever, 100 percent productive during an 8-hour day. An actual "pure" productive workday is approximately 60 to 80 percent of an 8-hour workday, because time may be taken up with staff meetings, phone calls, e-mails, and water cooler talks. Also, a long chain of 1-day activities may be assigned to one employee who is assumed to never get sick or take vacation. It is important that activity durations remain realistic. Durations should not be broken up simply to meet an artificial guideline. If the work required for the activity is estimated to extend beyond the reporting period, or if network logic dictates an activity duration longer than the reporting period for some other reason, then the activity duration should reflect this reality.

Certain activities within schedules naturally span more than the number of working days in the reporting period. For summary-level schedules, often created before detailed engineering is complete, durations might be longer than 1 or 2 months and lags might be more common. Within detailed schedules, LOE activities such as management and other oversight activities depend on the duration of the underlying discrete effort, so they span complete phases or even the entire project. LOE activities should be clearly marked in the schedule and should never appear on a critical path.

In some circumstances, it may be beneficial to use long-duration activities in a schedule to reduce complexity. For example, if 30 units of some item need to be constructed and each item has 15 individual steps, the complexity of the schedule can be reduced by creating 30 construction activities rather than 450 step activities. To ensure that long-duration activities can be effectively progressed, they should be monitored using incremental milestones. Incremental milestones—also called inch-stones—are used to track the completion of a long-duration activity. Incremental milestones should represent objective, product-oriented progress on the task and should be managed under a control process. They should enhance the performance visibility of the activity rather than represent arbitrary points in time. Inch-stones used to calculate performance for long-duration activities are sometimes referred to as quantifiable backup data, or QBD.

In addition, planning packages representing summarized or less-defined future work can be several months long. However, the duration of the planning packages must be estimated and they should still be integrated into network logic.

Finally, all activity durations in the schedule should be defined by the same time unit (hours, days, weeks) to facilitate calculating and monitoring the critical path. The day is the preferred time unit.

ESTIMATING DURATIONS

Activity durations should be realistic to ensure that forecasted program delivery dates and critical paths are reliable. Activity durations are often mistakenly determined solely by the time available to complete the program. However, activity durations should be based on the effort required to complete the activity, the resources available, and resource efficiency or productivity. This ensures that the dates in the schedule are determined by logic and durations rather than by wishful thinking or estimates that are constructed to meet a particular finish date objective.

If the estimated durations and supporting schedule network logic do not support the target deliverable date, then the program manager and the teams must discuss how to realistically compress the schedule, perhaps by adding more resources, adjusting scope, or setting a later finish date.[21]

Activity durations should be estimated under most likely or "normal" conditions, not under optimal, "success-oriented" conditions or with padded durations. Most likely conditions for estimated durations imply that duration estimates do not contain padding or margin for risk. Rather, risk margin should be introduced as separate schedule contingency activities to facilitate proper monitoring by management, as discussed in Best Practice 8. Durations also should not be unrealistically short or arbitrarily reduced by management to meet a program challenge.

Activity owners should be responsible for estimating durations for their activities. Activity owners may have experience from similar activities on past projects that can help them estimate the duration of the current activity. If the scheduler creates estimates, it is important that the underlying assumptions and durations are acceptable to the activity owners.

---

[21] Usually biased estimating has the purpose of achieving an earlier project finish date, often to satisfy management or the customer. That is, the direction of the estimating bias is typically not symmetrical.

All assumptions related to activity duration estimates should be documented in appropriate detail, such as a record of the methodology used to create the estimate (for example, parametric analysis of historical data or analogy to similar effort) and all supporting historical or analogous data. Documenting the basis for duration facilitates the communication of expectations between activity owners and decision makers and helps in estimating future analogous activity durations.

Finally, activity duration estimates for a WBS element in the schedule should clearly map to and correspond with the basis of the cost estimate for the same WBS element. For example, assumptions for the number of FTE workers underlying the cost estimate for a WBS element should also underlie the duration estimates for the WBS element in the schedule. This mapping need not necessarily be done at the lowest task level. However, at some level of the schedule, duration estimates should be supported by the basis of estimate.

As discussed in Best Practice 1 and Best Practice 3, a comprehensive IMS reflects all the activities of a program yet incorporates different levels of detail, depending on the information available at any point in time. Detailed activities reflect near-term, well-defined effort while less well-defined effort beyond the near term is represented within the schedule as planning packages. Resource assignments and duration estimates vary in detail according to how detailed activities are in the schedule.

Duration estimates for near-term detail activities should be related to the amount of work required, specific resource availability, and resource productivity. Long-term planning packages naturally have less-accurate resource availability and productivity information from which to estimate durations because they can be several months or years long. Estimates for the durations of planning packages are most likely to be based on analogies to historical projects, planners' experience, or standard productivity rates.

CALENDARS

Calendars in schedules specify valid working times for resources and activities. Resources can be assigned to calendars to define their availability. The availability of a resource in turn affects the dates and elapsed duration of the activity to which it is assigned. Activities should be directly tied to task calendars, which will define the valid times an activity can be worked. Calendars are defined by the number of work days per work week as well as the number of hours available each work day. Special nonwork days, known as exceptions, are defined in resource and activity calendars. Holidays and plant shutdown periods are examples of exceptions at the activity calendar level. Resource managers are responsible for identifying and assigning the correct calendar to their resources.

As is described in Best Practice 2, the proper use of resource and task calendars usually precludes the need for soft constraints in schedules. For example, a SNET constraint may be used to prevent a testing activity from beginning too soon while the testing facility is in use for an unrelated project. Instead of a SNET constraint, the test facility should be defined as a resource within the schedule and assigned to a calendar whose exceptions represent days other projects are using the facility. If the testing facility becomes available sooner or later than originally planned, the scheduler need only update the exceptions within the testing facility resource calendar. Task calendars may also be employed to represent a suspension of effort. For example, if severe weather requires the suspension of pouring concrete, the activity can be assigned to a specific calendar that prevents the work from being performed on certain days.

Another use for a resource calendar might be to exclude seasonal days from work. Outdoor construction probably cannot be conducted when the ground is frozen or rain is intense. Such an activity could start before the rain or the freeze but would then have to continue after the end of the exception. A calendar that excludes November through February for bad weather could be assigned to all outdoor construction activities. Such dates are well known from many years of experience. Defining resource calendars in this way allows for properly scheduling activities automatically according to network logic. Automatically updating the schedule gives greater confidence in float calculations and the derived critical paths. Resource calendars must also be adjusted for planned over-time—to allow a resource to work in an otherwise nonworking period.

While resource calendars allow for greater insight into resource availability, having too many calendars may interfere with critical path analysis because calendars can affect float calculations. The benefit of using resource calendars to track exceptions for individual resources should be tempered with their possibly negative effects on the critical path (see Best Practice 6 for more information). In addition, the administration necessary to build and maintain many resource calendars may quickly outweigh their benefits.

Program managers must ensure that calendars are properly defined because schedules can incorrectly represent the forecasted start, finish, and durations of planned work if resources are assigned an incorrect calendar. A common mistake allows all activities within a schedule to simply adhere to the default calendar within the scheduling software. However, a default calendar rarely has national holidays appropriately defined as exceptions and does not define specific blackout periods or related exceptions. Similarly, the general project calendar that would have excluded holidays still may not represent the work practices of all resources. For instance, a testing facility may work 24 hours a day while some personnel work 4 10-hour days a week.

Additionally, if management has planned that work will be performed 7 days a week within the schedule, it is crucial that the people assigned activities are aware of the schedule. Planning effort can prevent unexpected delays in the project and the unnecessary use of schedule contingency. Establishing realistic calendars provides for greater accuracy of dates and may reveal opportunities to advance the work. Case study 9 shows how incorrect calendars can affect planned dates.

Case Study 9: The Effect of Incorrect Calendars, from *Transportation Worker Identification Credential*, GAO-10-43

The pilot schedule for the Department of Homeland Security's Transportation Worker Identification Credential (TWIC) program included duration estimates for all activities, but GAO could not be certain of their reliability. Nearly 86 percent (259 of 302) of the activities identified in the schedule were assigned to a 7-day calendar that did not account for holidays. While pilot sites may normally operate on a 7-day schedule, resources for conducting pilot activities such as installing readers and associated infrastructure such as cables and computers or analyzing the results of pilot data may not be available on weekends. By using a 7-day calendar, the schedule inaccurately indicated that approximately 28 percent more workdays were available each year than actually were available. In addition, our analysis of an earlier TWIC schedule found that calendars did not include the appropriate exceptions for holidays. For example, multiple pilot sites were scheduled to finish submitting pilot test data on Christmas Day.

Best practices in project management include obtaining stakeholders' agreement with project plans, such as the schedule. Because the schedule was not shared with the individual pilot sites, responsible pilot officials had no opportunity to comment on whether the 7-day schedule matched available resources. Therefore, pilot participants may not have the resources, such as employees who can work on weekends, to meet pilot goals.

If an activity is defined as taking 60 days, or approximately 2 months using a 7-day calendar, the reality may be that participants work a 5-day workweek with the result that the activity takes approximately 3 months to complete—1 month longer than scheduled.

GAO, *DHS Transportation Worker Identification Credential: Progress Made in Enrolling Workers and Activating Credentials but Evaluation Plan Needed to Help Inform the Implementation of Card Readers*, GAO-10-43 (Washington, D.C.: November 18, 2009).

A best practice when considering the duration of activities around the U.S. holiday season (Thanksgiving through New Year's Day) is to recognize that productivity is generally low then and that many workers take extended holidays during this time.[22]

BEST PRACTICES CHECKLIST: ESTABLISHING THE DURATION OF ALL ACTIVITIES

- Activity durations are directly related to the assigned resources and estimated work required.

- In general, estimated detailed activity durations are shorter than the reporting period management requires.

---

[22]Schedule risk analysis can handle this phenomenon with probabilistic calendars, where any day has less than a 100 percent chance of being a working day, with that percentage reflecting the experience of people taking time off during a holiday season.

- Durations are as short as possible, to a point, to facilitate the objective measurement of accomplished effort.

- Long durations should be broken into shorter activities if logical breaks can be identified in the work being performed. If it is not practical to divide the work into smaller activities or insert intermediate milestones, justification for long durations is provided in the schedule basis document.

- Very short durations, such as 1 day or less, may imply a schedule that is too detailed and require more-frequent updates to schedule duration and logic than is otherwise necessary.

- LOE activities are clearly marked in the schedule and derive their durations from other discrete activities.

- All activity durations within the schedule are defined by the same time unit (hours, days, weeks). Days are preferred.

- Planning packages representing summarized or less-defined future work should be integrated into network logic.

- Activity durations are estimated under most likely conditions, not optimal or "success-oriented" conditions. "Most likely" for estimated durations implies that duration estimates do not contain padding or margin for risk. They should also not be unrealistically short or arbitrarily reduced by management to meet a program challenge.

- All assumptions related to activity duration estimates are documented in appropriate detail, such as describing the methodology used to create the estimate (for example, parametric analysis of historical data or opinion of a subject matter expert) and all specifying supporting historical or analogous data.

- Activity duration estimates for a WBS element in a schedule should clearly map to and correspond with the basis of the cost estimate for the same WBS element.

- Calendars are used to specify valid working times for activities and, when feasible, resources.

BEST PRACTICE 5

Verifying That the Schedule Can Be Traced Horizontally and Vertically

Best Practice 5: The schedule should be horizontally traceable, meaning that it should link products and outcomes associated with other sequenced activities. Such links are commonly referred to as "hand-offs" and serve to verify that activities are arranged in the right order for achieving aggregated products or outcomes. The schedule should also be vertically traceable—that is, varying levels of activities and supporting subactivities can be traced. Such mapping or alignment of levels enables different groups to work to the same master schedule.

Horizontal traceability demonstrates that the overall schedule is rational, has been planned in a logical sequence, accounts for the interdependence of detailed activities and planning packages, and provides a way to evaluate current status. Schedules that are horizontally traceable depict logical relationships between different program elements and product handoffs. Horizontally traceable schedules support the calculation of activity and milestone dates and the identification of critical and near-critical paths.

Horizontal traceability applies to both an individual project schedule and the entire IMS, which may consist of multiple files. A horizontally traceable IMS includes complete logic from program start to program finish and fully integrates the entire scope of work from all parties in the program. Horizontal traceability ensures that forecasted dates within the schedule will be determined by network logic and progress to date rather than by artificial constraints. Any logic errors in the summary, intermediate, and detailed schedules will make dates between schedules inconsistent and will cause managers and activity owners to differ in their expectations.

Detail activities, milestones, and planning packages in a horizontally traceable schedule are linked to one another, preferably through straightforward finish-to-start logic at the detailed level that represents the required inputs and outputs in a planned effort. Milestones representing key decisions or deliverables should have each predecessor activity traced and validated to make certain that they are directly related to accomplishing the milestone.

In particular, "giver/receiver" (G/R) milestones, common when multiple schedules are linked to form the IMS, must be clearly identified and logically linked between schedules. Giver/receiver milestones represent dependencies between schedules, such as hand-offs between integrated product teams and delivery and acceptance of government-furnished equipment. For example, a production schedule may include "receiver" milestones from outside suppliers representing the delivery of material and "giver" milestones representing the delivery of the produced article to the testing team. Likewise, the test schedule will include a receiver milestone that represents the receipt of the production article. Key G/R milestones should be defined in the schedule basis document.

A horizontally traceable schedule dynamically reforecasts the date of a key milestone through network logic if activities related to accomplishing it are delayed longer than scheduled. For example, if the duration of a key milestone's predecessor activity is greatly extended relative to available float, the date of the key milestone should slip. Activities whose durations are extended many days but have no effect on key milestones should be examined for unrealistic or dangling logic. Horizontal traceability is directly related to Best Practice 2 and is the result of a dynamic IMS whose activities are properly sequenced.

VERTICAL TRACEABILITY

Vertical traceability demonstrates the consistency of dates, status, and scope requirements between different levels of a schedule—summary, intermediate, and detailed. When schedules are vertically traceable, lower-level schedules are clearly consistent with upper-level schedule milestones, allowing for total schedule integrity and enabling different teams to work to the same schedule expectations. In this way, management can base informed decisions on forecasted dates that are reliably predicted in detailed schedules through network logic and actual progress.

In addition, vertical traceability allows managers to understand the effect on key program and G/R milestones if their lower-level activities are delayed. An activity owner should be able to trace activities to higher-level milestones within intermediate and summary schedules. Even though their activities may be rolled into a higher-level milestone, responsible owners should be able to identify when and how their product affects the program.

All levels of schedule data, from detailed through summary schedules, should be derived from the same IMS. Ideally, the same schedule serves as the summary, intermediate, and detailed schedule by simply creating a summary view filtered on summary activities or higher-level WBS milestones. Summary schedules created by rolling up the dates and durations of lower-level elements are inherently vertically integrated.

Often the program schedule is represented by presentation software in a completely different format. If alternative presentations represent the program schedule, the program management office should be able to demonstrate that they are consistent with the schedule.

It is important to note that vertical traceability is not simply the ability to collapse WBS elements within the same contractor schedule. Vertical traceability implies the ability of the contractor schedule to roll up into the overall program schedule, which includes government activities, other contractor schedules, and interfaces with external parties, even if constructed in a separate computer file or different software package. All the program's project schedule information, from the suppliers through the contractors and government agencies, should be collapsible into one overall IMS.

Finally, vertical traceability applies to all schedule data that are reported to and by program management. That is, schedule information such as forecasted dates, predecessor logic, and critical path activities reported to management should be rooted in and traceable to the actual program IMS. Case study 10 gives an example of the absence of vertical traceability between schedules.

Case Study 10: Missing Vertical Traceability, from *DOD Business Systems Modernization*, GAO-14-152

For this study, GAO reviewed the most current schedule and cost estimates that supported DOD's February 2012 Milestone B decision, which determined that investment in the Air Force Defense Enterprise Accounting and Management System was justified. Our analysis did not find vertical traceability within the schedule for Releases 1 and 2 of Increment 1—the ability to consistently trace work breakdown structure activities between detailed, intermediate, and master schedules.

For example, we traced three activities between the government schedule and the underlying prime contractor schedule, and in each case we found mismatching start dates that differed by a day, a week, and a month. Vertical traceability ensures that representations of the schedule to different audiences are consistent and accurate. Unless the schedule is vertically traceable, lower-level schedules will not be consistent with upper-level schedule milestones, affecting the integrity of the entire schedule and the ability of different teams to work to the same schedule expectations.

GAO, *DOD Business Systems Modernization: Air Force Business System Schedule and Cost Estimates,* GAO-14-152 (Washington, D.C.: February 7, 2014).

BEST PRACTICES CHECKLIST: VERIFYING THAT THE SCHEDULE CAN BE TRACED HORIZONTALLY AND VERTICALLY

- The schedule is horizontally traceable. That is, the schedule

  o depicts logical relationships between different program elements and product hand-offs and clearly shows when major deliverables and hand-offs are expected;

  o includes complete logic from program start to program finish and fully integrates the entire scope of work from all involved in the program;

- includes milestones representing key decisions or deliverables with traced and validated predecessor activities to ensure that they are directly related to completing the milestone;
- clearly identifies and logically links giver/receiver milestones between schedules that are defined in the schedule basis document;
- dynamically reforecasts the date of a key milestone through network logic if activities related to accomplishing the milestone are delayed;
- is affected by activities whose durations are extended by many days.

- The schedule is vertically traceable. That is, it

  - demonstrates that data are consistent between summary, intermediate, and detailed levels including dates that are frequently validated through a process;
  - allows activity owners to trace activities to higher-level milestones with intermediate and summary schedules;
  - allows lower-level schedules to be rolled up into the overall program schedule, which includes government activities, other contractor schedules, and interfaces with external parties.

Best Practice 5: Verifying That the Schedule Can Be
Traced Horizontally and Vertically

BEST PRACTICE 6

Confirming That the Critical Path Is Valid

Best Practice 6: The schedule should identify the program's critical path—the path of longest duration through the sequence of activities. Establishing a valid critical path is necessary for examining the effects of any activity's slipping along this path. The program's critical path determines the program's earliest completion date and focuses the team's energy and management's attention on the activities that will lead to the project's success.

The critical path is generally defined as the longest continuous sequence of activities in a schedule. It defines the program's earliest completion date or minimum duration. Activities on this path are termed "critical path activities." Typically, the sequence of activities with the longest total duration is also the path through the network with the lowest total float. As Best Practice 7 shows, total float is the time an activity can slip before its delay affects the program end date. When the network is free of date constraints, critical activities have zero float, and therefore any delay in the critical activity causes the same day-for-day delay in the program forecast finish date.

For example, if an activity on the critical path is delayed by a week, the program finish date will be delayed by a week unless the slip is mitigated. Therefore, the critical path is most useful as a tool to help determine which activities deserve focus and, potentially, management help. The critical path assists program management in prioritizing resources to have the most positive effect on program performance.[23]

In figure 28, the critical path is made up of three detail work activities: "install carpeting and wood flooring," "install kitchen cabinets and countertops," and "set plumbing fixtures." The activities are 6 working days, 2 working days, and 3 working days long, respectively. The minimum duration of this particular sequence of activities is therefore 11 days. An additional day's activity, "install laundry room cabinets and countertops," is performed in parallel with "install kitchen cabinets and countertops"—that is, the installation of cabinets and countertops in both the laundry room and the kitchen may start after the installation of the carpeting and wood flooring finishes.

---

[23]In this section, we discuss the *deterministic* critical path—that is, the path as defined by the initial or current set of inputs in the schedule model. However, the true critical path of a schedule is uncertain because durations of activities are uncertain. Best Practice 8 discusses a *probabilistic* or *risk* critical path that is based on assumptions about estimating error and risk.

Figure 28: The Critical Path and Total Float

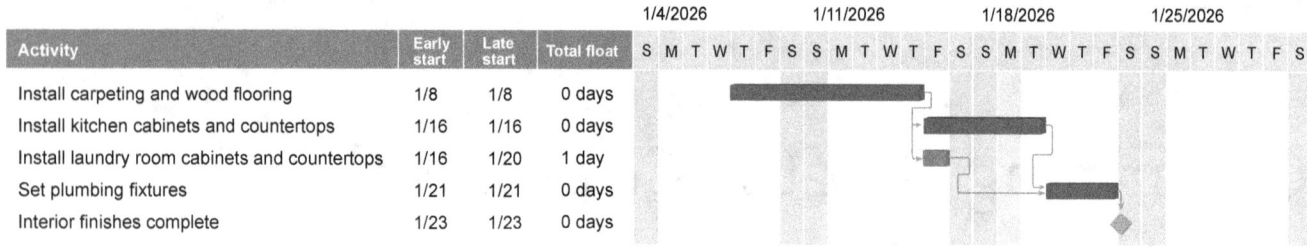

Activity	Early start	Late start	Total float
Install carpeting and wood flooring	1/8	1/8	0 days
Install kitchen cabinets and countertops	1/16	1/16	0 days
Install laundry room cabinets and countertops	1/16	1/20	1 day
Set plumbing fixtures	1/21	1/21	0 days
Interior finishes complete	1/23	1/23	0 days

Source: GAO. | GAO-16-89G

When cabinets and countertops have been installed in the kitchen and laundry room, the plumbing fixtures can be set, and then the interior finishing of the house is considered complete. Because "install laundry room cabinets and countertops" is 1 day shorter than "install kitchen cabinets and countertops," it does not directly affect the start date of the "interior finishes complete" milestone. Notice that "install laundry room cabinets and countertops" has an early start of January 16 and a late start of January 20. In other words, laundry room cabinet and countertop installation could start as late as January 20 (1 working day later than planned) and, if finished in 1 day, would have no effect on the "interior finishes complete" milestone date of January 23.

The difference between early and late dates for "install laundry room cabinets and countertops" yields 1 working day of total float. The early and late start dates for "install kitchen cabinets and countertops," however, are equal. If the kitchen cabinet installation is delayed by 1 day, or if its duration extends by 1 day, then the "interior finishes complete" milestone slips by 1 day. "Install kitchen cabinets and countertops" has zero float and is on the critical path.

Scheduling software automatically calculates a critical path through a network of activities by defining as critical the activities that have less than predefined total float. Typically, total float is set to zero, and the scheduling software marks as critical all activities with zero or less-than-zero total float.

Activities with total float within a narrow range of the critical path total float are "near-critical" because they can quickly become critical if their float is used up in a delay. Near-critical paths need only a small extension of time to become critical. Management must monitor critical and near-critical activities through sound schedule management because any delay in them will delay the entire program. Near-critical paths are monitored according to a float threshold tailored to the program. For example, a brief schedule might consider a 5-day slip to be a near-critical threshold. In programs scheduled to take years, a 2- or 3-month's slip in near-critical paths might make the path critical. Because prolonging a schedule by 5 days on a short project is as easily possible as prolonging a multiyear project several months, program managers should manage all near-critical and critical paths.

The critical path is not constant. The sequence of activities that make up the critical path changes as activities are delayed, finished early, occur out of planned sequence, and so on. Activities that were previously critical may become noncritical, and activities that were not critical may become critical.

It is crucial that program management understand that an important activity is not necessarily "critical." At any point in time, the critical path may or may not contain activities that management believes are particularly important. A delay in an activity may be important for any number of reasons related to scope and cost without delaying the finish milestone date. In contrast, some mundane activities—training, for example— may be on the critical path and not particularly risky but can delay the program finish date if they take longer to accomplish. Similarly, an activity of long duration should not be referred to as a "critical path activity" simply because it will take a long time to accomplish. "Critical activity" in scheduling parlance has a specific definition that should be adhered to when reporting and evaluating schedule data.

The Critical Path and the Longest Path

The critical path is theoretically the sequence of activities that represents the longest path between the program's start and finish dates. If the program has started, then the critical path will extend from the program's current status date to the program's forecasted finish date. In reality, however, as a schedule becomes more complex, total float values may not necessarily represent a true picture of the number of days an activity can slip. For example, multiple calendars, out-of-sequence progress, date constraints, and leveled resources can all produce misleading values of total float in complex schedules, leading to a misrepresentation of the sequence of activities that actually drives the program finish date.

As we noted in Best Practice 2, date constraints may cause activities to become critical, regardless of the total float that may be available if not constrained in the network. Specifically, backward-pass date constraints on activities other than the finish milestone will influence the criticality of activities. Hence, where constraints are many, there may be many more activities with zero or negative total float than activities that are actually driving the key program completion milestone.[24]

Figure 29 shows in the construction schedule two sequences of activity necessary to complete interior rough-in: critical activities (in red) and noncritical activities (in blue). The critical path is also the sequence of activities that represents the longest path to the "interior rough-in complete" milestone. The sum of the durations of the critical activities is 13 days. That is, the minimum duration from the start date of "install exterior doors and windows" to the "interior rough-in complete" milestone is 13 working days. A second path of activities, consisting of "rough in interior HVAC and through-roof

---

[24]Negative float and its causes are discussed in detail in Best Practice 7.

penetration" and "inspect rough-in HVAC," after the installation of the exterior doors and windows, has 2 days of available float and is therefore not considered critical.

Figure 29: The Critical Path and the Longest Path

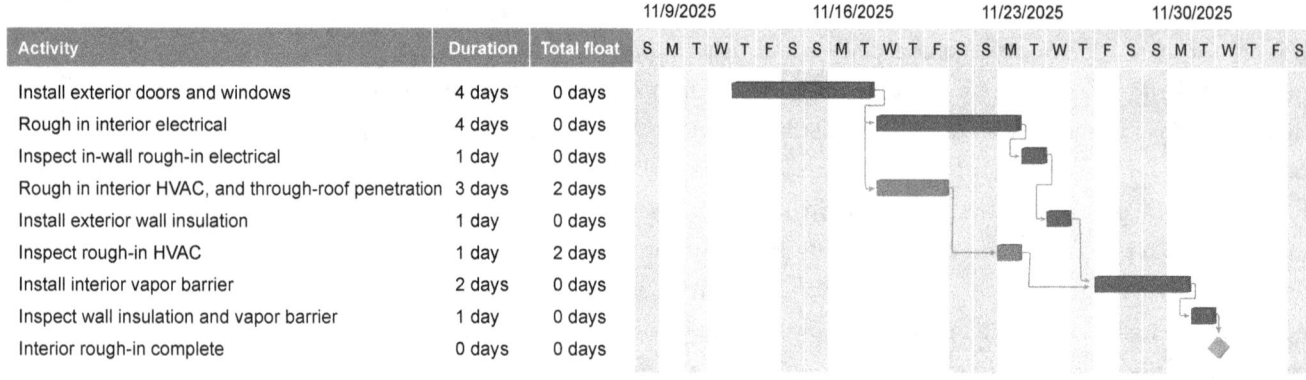

Activity	Duration	Total float
Install exterior doors and windows	4 days	0 days
Rough in interior electrical	4 days	0 days
Inspect in-wall rough-in electrical	1 day	0 days
Rough in interior HVAC, and through-roof penetration	3 days	2 days
Install exterior wall insulation	1 day	0 days
Inspect rough-in HVAC	1 day	2 days
Install interior vapor barrier	2 days	0 days
Inspect wall insulation and vapor barrier	1 day	0 days
Interior rough-in complete	0 days	0 days

Source: GAO. | GAO-16-89G

If the general contractor were to introduce a backward-pass date constraint on "inspect rough-in HVAC"—directing, for example, that the inspection be performed on November 24—then the date constraint on "inspect rough-in HVAC" would consume that path's entire total float, redefining it as critical according to the total float criticality threshold of zero (figure 30).

Figure 30: Critical Path Activities Not on the Longest Path

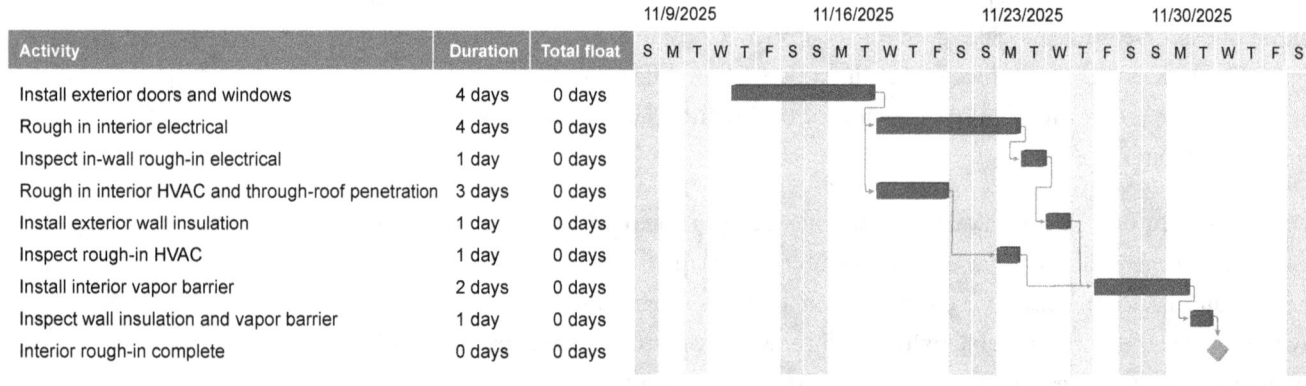

Activity	Duration	Total float
Install exterior doors and windows	4 days	0 days
Rough in interior electrical	4 days	0 days
Inspect in-wall rough-in electrical	1 day	0 days
Rough in interior HVAC and through-roof penetration	3 days	0 days
Install exterior wall insulation	1 day	0 days
Inspect rough-in HVAC	1 day	0 days
Install interior vapor barrier	2 days	0 days
Inspect wall insulation and vapor barrier	1 day	0 days
Interior rough-in complete	0 days	0 days

Source: GAO. | GAO-16-89G

The network now has 2 more critical activities than the network without the date constraint and shows 2 parallel critical paths in terms of total float. However, the longest path—in terms of duration—remains the same, regardless of the date constraint. While "inspect rough-in HVAC" is marked critical in the schedule, it actually has 2 days of relative float because it can slip 2 days before causing the "interior finishes complete" milestone to slip. The longest path is also referred to as the driving path because it determines the date of the key milestone. A driving path can be identified for any key milestone or activity to determine the sequence of activities driving its finish date.

When the critical path is not the longest path, the longest path is preferred because it represents the activities that are driving the sequence of start dates directly affecting the estimated finish date, if we ignored the presence of date constraints. Therefore, rather than simply filtering on activities that are marked critical by the scheduling software, management should be aware of the activity "drivers" that are determining the schedule finish date.

Moreover, driver activities may or may not have the lowest total float values when activities other than the program finish milestone have date constraints. Continuing with the framing example in figure 30, suppose the general contractor mandates the inspection of the rough-in HVAC by Thursday, November 20 (figure 31).

Figure 31: The Longest Path and the Lowest-Float Path

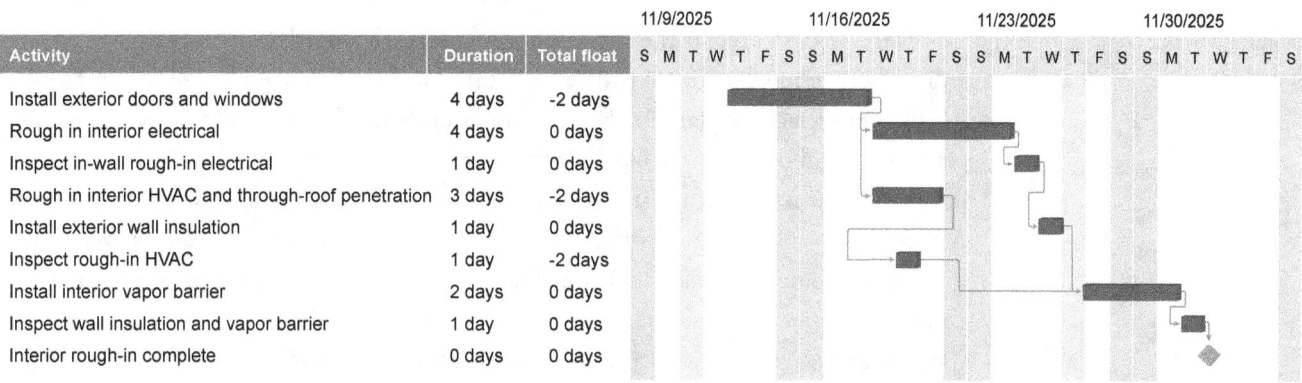

Activity	Duration	Total float
Install exterior doors and windows	4 days	-2 days
Rough in interior electrical	4 days	0 days
Inspect in-wall rough-in electrical	1 day	0 days
Rough in interior HVAC and through-roof penetration	3 days	-2 days
Install exterior wall insulation	1 day	0 days
Inspect rough-in HVAC	1 day	-2 days
Install interior vapor barrier	2 days	0 days
Inspect wall insulation and vapor barrier	1 day	0 days
Interior rough-in complete	0 days	0 days

Source: GAO. | GAO-16-89G

The lowest total float path consists now of "install exterior doors and windows," "rough in interior HVAC and through-roof penetration," and "inspect rough-in HVAC." However, neither "rough in interior HVAC and through-roof penetration" nor "inspect rough-in HVAC" is determining the finish date of the "interior rough-in complete" milestone. Most scheduling software calculates activity drivers along with critical activities.

COMMON BARRIERS TO A VALID CRITICAL PATH

As noted above, the critical path ideally represents the longest path, as when the schedule network is free of backward-pass constraints and activities on this path have the least float in the network. In this section, we highlight issues that prevent the critical path from being the longest path. When these issues arise, it is imperative that management recognize not only critical path activities—that is, activities with the lowest total float— but also activities that are truly driving the finish date of key milestones.

Calculating a critical path is directly related to the logical sequencing of activities. Missing or convoluted logic and artificial date constraints prevent the calculation of a valid critical path; they can cause activities that are not critical to appear to be critical.

Successfully identifying the critical path relies on a valid, reliable schedule. This includes capturing all activities (Best Practice 1), proper sequencing of activities (Best Practice 2), horizontal traceability (Best Practice 5), the reasonableness of float (Best Practice 7), accurate status updates (Best Practice 9), and—if there are resource limitations—assigning resources (Best Practice 3).

It is essential that the critical path be evaluated before the schedule is baselined and after every status update to ensure that it is valid. If the schedule is missing activities, then the critical path will not be valid. Moreover, if the critical path is missing dependencies or has date constraints, lags, or LOE activities or it is not a continuous path from the current status date to the finish milestone, then it is not valid.

Continuous through All Activities

Unless the IMS represents the entire scope of effort, the scheduling software will report an incorrect or invalid critical path. As discussed in Best Practice 1 and Best Practice 4, the IMS must include planning for all activities that have to be accomplished in the program. A critical path for one block, increment, or contract for a multiyear multiphased program is not a sufficient plan that can reliably forecast the finish date for the program. A critical path should exist for the entire program because detail activities, as well as long-term planning packages, must be logically linked within the schedule to create a complete picture of the program from start to finish. In addition, for projects under way, the longest path or critical path should start with at least one in-progress activity that has an actual start date.[25]

The critical path should be a continuous sequence of activities from the schedule status date to the finish milestone. In general, the sequence of activities should have no breaks and no large gaps of unaccounted time. The critical path may branch off into several sequences of activities, but they must ultimately converge at the finish milestone. Sorting the schedule by activity start date, filtering by critical activities, and visually assessing the sequence of activities in a Gantt chart is an easy way to assess the practicality of the calculated critical path. Ideally, the Gantt chart displays a continuous waterfall of activities from the status date to the program finish date that are logically linked with finish-to-start relationships. Case study 11 shows why program-wide critical paths are important.

---

[25] As noted in Best Practice 9, in principle, a critical activity could be scheduled to start the next day after a status update. It would therefore not be in progress at that time, although it would be scheduled to start as soon as possible.

Officials from FAA's Collaborative Air Traffic Management Technologies system said that because of its 6-month spiral development, the program schedule could not deliver a single critical path for the entire program. Instead, it had critical paths by release. To produce the critical paths, the prime contractor used a constraint on the key deliverable finish milestone for each release.

At the time of GAO's analysis, officials stated that only work related to Release 5 was subject to a critical path analysis because that release was management's current focus. GAO was able to trace a continuous critical path for Release 5, beginning with the project status date and ending with the Release 5 finish milestone.

However, the validity of the Release 5 critical path was hampered by seven in-progress and remaining detail activities within Release 5 that had over 1,300 working days of total float. GAO also determined that a small number of activities outside Release 5 were under way (for example, activities related to other task orders). Because these activities fell outside the Release 5 task order, and therefore outside the purview of a Release 5 critical path analysis, management may not have been fully aware of the effect of any delay in these activities.

GAO, *FAA's Acquisitions: Management Challenges Associated with Program Costs and Schedules Could Hinder NextGen*, GAO-12-223 (Washington, D.C.: February 16, 2012).

Breaks in the critical path should be examined immediately and justified or otherwise addressed. Common causes of noncontinuous critical paths are that

- the start or finish date of an activity is driven by a constraint;
- a successor activity is driven by an unexplained lag;
- the start date of an activity is driven by an external predecessor;
- activities are scheduled according to different calendars, as when a predecessor activity ends in a nonworking period for the successor; and
- resource leveling is causing delays.

For example, if an activity on the critical path starts some days or weeks after its driving predecessor finishes (assuming finish-to-start logic) because of a start date constraint or an unexplained lag, then the path is considered to be noncontinuous and broken. In each scenario above, additional total float may occur in the predecessor activities; it would break the sequence of activities with zero total float. The program management team must determine whether this additional total float is meaningful. If they decide that it is not, the threshold for total float criticality may have to be raised.

Noncontinuous critical paths are often caused by the use of multiple calendars. Once multiple calendars are introduced into a schedule network, total float calculations may be adversely affected. In the sequence of activities in figure 32, all activities and resources are assigned standard 5-day workweek calendars except "install footing and pier rebar."

The installation activity is assigned a 7-day workweek calendar as well as resources that can work on Saturdays and Sundays. Because Sunday is a valid workday for this activity, it has one day of total float available: its early start is Saturday, September 20, and its late start is Sunday, September 21. However, its successor, "inspect footing and pier rebar," can occur only on Monday through Friday, so its early and late start dates are Monday, September 22. The critical path (red activities), as defined by total float, is discontinuous at the weekend.

Figure 32: The Effect of Multiple Calendars on the Critical Path

Activity	Early start	Late start	Total float
Lay out and form footings and pier pads	9/17	9/17	0 days
Install footing and pier rebar	9/20	9/21	1 day
Inspect footing and pier rebar	9/22	9/22	0 days
Pour footings and pads	9/23	9/23	0 days
Cure footings and pads	9/24	9/24	0 days

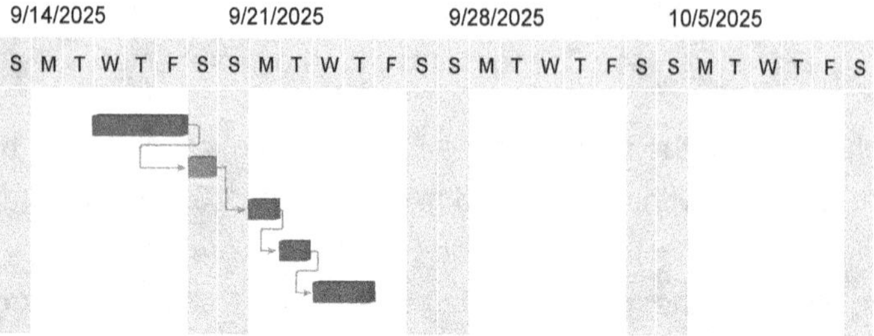

Source: GAO. | GAO-16-89G

Figure 32 is a simple instance that can quickly become more complex. For example, if "lay out and form footings and pier pads" is also assigned a 7-day workweek and 7-day resources, as a predecessor to "install footing and pier rebar," it too gains a day of total float.

As we noted in Best Practice 4, while the use of multiple calendars allows for greater insight into resource availability and precision of forecasted start and finish dates, their use must be tempered with their possibly negative effect on the critical path. The threshold for total float criticality may need to be raised to capture the entire critical path, which in turn complicates the tracking of near-critical paths. Because the longest path makes no reference to total float, it is the only guaranteed method of identifying the driving sequence of activities when using multiple calendars. But when the longest path is not easily traceable, schedulers may opt to simplify the network by avoiding multiple calendars, especially if very few activities need a slightly different calendar.

The Number of Critical Activities

In general, assessing the quality of the critical path by predetermining the number of activities that should be critical is not useful. The number of activities on the critical path depends on the visibility required to manage the program and reduce risk. However, if the ratio of critical path activities to the total remaining activity count is nearly 100 percent, then the schedule may be overly serial and resource limited. Conversely, if only a few activities are on the critical path and if all represent LOE, then the critical path is being driven by supporting effort and will not identify effort that is driving key milestones.

Logical Sequencing

Float calculations are directly related to the logical sequencing of events (see Best Practice 7). Because float dictates the criticality of activities, the critical path is directly related to the logical sequencing of events and float calculations. If activities are missing dependencies, linked incorrectly, or performed out of sequence, float estimates will be miscalculated. Incorrect float estimates will result in an invalid critical path, hindering management's ability to reallocate resources from noncritical activities to those that must be completed on time. Errors or incomplete logic often cause values of total float that do not represent the state of the program schedule (we discuss the effect of dependencies on total float in Best Practice 7 and describe out-of-sequence progress in Best Practice 9).

Date Constraints

We saw in Best Practice 2 that placing a hard constraint on an activity fixes the dates and immediately causes the activity to become critical. It is therefore possible to use hard constraints as a working tool while developing a schedule to calculate total available float up to key milestones. The temporary use of hard constraints is also valuable for assessing the likelihood that using available resources can achieve the planned activity date. However, using hard constraints simply to fix activity dates at certain points in time immediately convolutes critical path calculations. It also reduces the credibility of any schedule date on activities that logically occur after the hard constraint. In this case, the critical path is no longer the longest path; instead, each hard constraint in the schedule generates its own sequence of critical activities, and the purpose of CPM scheduling is defeated.

Lags

The critical path should be free of lags because they obfuscate the identification and management of critical activities. As described in Best Practice 2, a lag is used in a schedule to denote the passing of time between two activities. Lags cannot represent work and cannot be assigned resources. Lags simply delay the successor activity, and no effort or resources are associated with them.

Lags are often misused to force successor activities to begin on specific dates. Evaluating a critical path that includes lags can therefore be difficult because the lag may or may not represent work that has to be accomplished by some resource, either internal or external to the program. In addition, because lags are not represented in a schedule as an activity, they can easily be overlooked as drivers of the finish milestone date. A lag that indicates a waiting period, such as for document review or materials delivery, should be replaced with an activity that can be actively monitored, statused, and perhaps mitigated if necessary.

In figure 33, "inspect rough-in HVAC" now has a 2-day lag on its finish-to-start relationship to the activity "install interior vapor barrier," representing 2 days of paperwork that the general contractor's company must perform. The total duration of "inspect rough-in HVAC," including its lag, is 3 days. The activity becomes critical because the 2-day lag consumes its 2 days of total float. A report to the general contractor listing the current critical activities in the project would list "inspect rough-in HVAC" as critical but would more than likely fail to mention that its successor lag is causing it to be on the critical path, not the inspection activity itself. Any management initiatives by the general contractor to reduce the criticality of the inspection activity would fail, because the initiatives would be misdirected. The planned effort that is causing the criticality is concealed by the lag.

Figure 33: The Critical Path and Lags

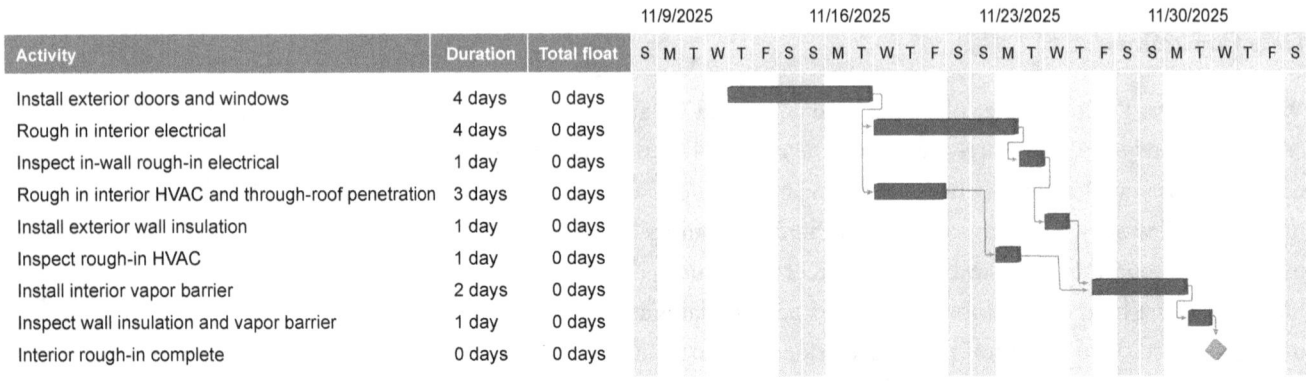

Source: GAO. | GAO-16-89G

Level-of-Effort and Other Long-Duration Activities

We learned in Best Practice 1 that some schedulers represent LOE work by activities that have estimated durations rather than by hammock activities, the preferred practice. Their long durations mean that these activities may inadvertently define the length of a project and, thus, become critical if they are linked to successors through logic. A critical path cannot include LOE activities because, by their very nature, they do not represent discrete effort. The duration of LOE activities is determined by the overall duration of the discrete work they support. Therefore, an LOE activity cannot drive any successor and become critical. The logic should be placed on the detailed activities that the LOE resources support.

In the house construction framing example in figure 34, we see an overarching general contractor project management activity planned through November 12 to ensure that all framing activities are managed properly and all documentation is created and filed appropriately. However, the LOE project management activity now determines the minimum duration of the framing sequence of activities, rather than the sequence of actual work that has to be accomplished to complete framing. It is impossible to reduce the duration of framing by adding resources: adding resources to a level-of-effort activity simply increases the work. Using this schedule, management has no indication of which activities can slip and which will respond positively to additional resources and reduce the risk of finishing late.

Figure 34: An Incorrect Critical Path with Level-of-Effort Activities

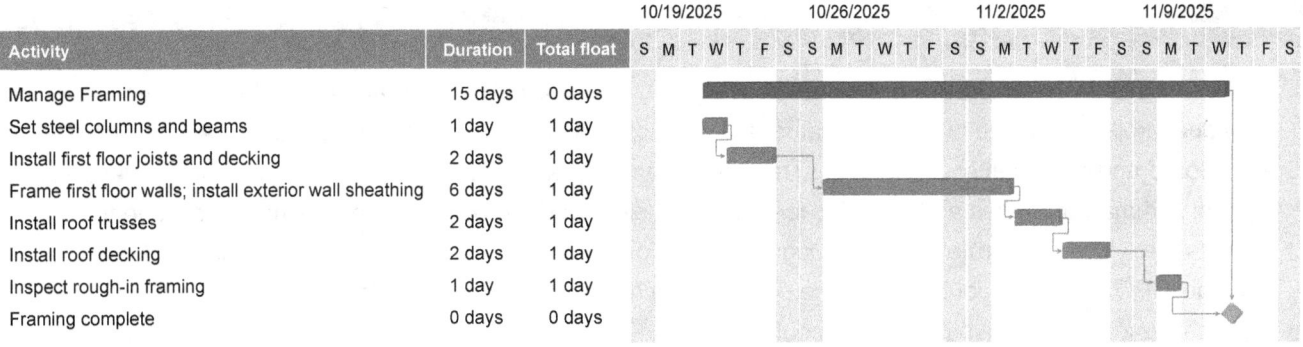

Activity	Duration	Total float
Manage Framing	15 days	0 days
Set steel columns and beams	1 day	1 day
Install first floor joists and decking	2 days	1 day
Frame first floor walls; install exterior wall sheathing	6 days	1 day
Install roof trusses	2 days	1 day
Install roof decking	2 days	1 day
Inspect rough-in framing	1 day	1 day
Framing complete	0 days	0 days

Source: GAO. | GAO-16-89G

Long-duration non-LOE activities in the detail planning period should be reevaluated to determine if they can be broken down into more manageable pieces, particularly if they appear on the critical path. If work is broken into smaller pieces, some portions of it might be able to start earlier or in parallel with other portions or might be reassigned to other available resources, saving time and possibly moving the activity off the critical path. In particular, summary activities should never be on the critical path. As described in Best Practice 2, linking summary activities obfuscates the logic of the schedule. Tracing logic through summary links does not impart to management the sequence in which lower-level activities should be carried out. If a summary activity becomes critical, there is no way to determine which subactivities are critical and which are not.

Predetermined Critical Activities

Finally, because all activities can become critical under some circumstances, intermediate and summary-level schedules should derive their critical paths from detailed schedules while recognizing that this will not be as informative as the full critical path. For example, separate summary schedules should not be created by selectively choosing lower-level milestones and detail activities that management has determined are important to monitor. Activities may be important in terms of cost or scope, but there is little guarantee that hand-picked activities will determine the finish milestone date as the program progresses. By only reporting on or monitoring the progress of favorite activities, management risks losing sight of previously unimportant activities that may now be driving the program's finish date (see case study 12).

The U.S. Citizenship and Immigration Services (USCIS) Transformation Program schedule consisted of 18 individual project schedules. To provide an alternative to an IMS and to ease reporting to the Acquisition Review Board and other senior officials, the Transformation Program Office (TPO) developed a high-level tracking tool summarizing dates and activities for the first release of the program based on individual schedules such as the Office of Information Technology and solutions architect schedule, which TPO did not directly manage. TPO used this tool to ensure the coordination and alignment of activities by collaborating with staff responsible for managing individual schedules.

However, this tracking tool was not an IMS because it did not integrate all activities necessary to meet the milestones for Release A. Rather, it was a selection of key activities drawn from the individual schedules that USCIS components and the solutions architect maintained. Moreover, the Transformation Program Manager expressed concern in a May 2011 program management review that the information reported in the high-level tracking tool was not being reported in the individual schedules. In addition, this tracking tool was not an IMS because it did not show activities over the life of the program. That is, it had no dates or activities for the four other releases' set of work activities or information on how long they would take and how they were related to one another.

As a result, program officials would have difficulty predicting, with any degree of confidence, how long completing all five releases of the Transformation Program would take. Program officials would also have difficulty managing and measuring progress in executing the work needed to deliver the program, thus increasing the risk of cost, schedule, and performance shortfalls. Last, USCIS would be hindered in accurately communicating the status of Transformation Program efforts to employees, the Congress, and the public.

In addition, neither of the two underlying schedules that GAO assessed had a valid critical path. The USCIS Office of Information Technology schedule had missing or dangling logic on over 60 percent of its remaining activities. The solutions architect's schedule was missing logic in 40 percent of its remaining activities and contained excessive constraints, lags, and dangling logic.

GAO, *Immigration Benefits: Consistent Adherence to DHS's Acquisition Policy Could Help Improve Transformation Program Outcomes*, GAO-12-66 (Washington, D.C.: November 22, 2011).

RESOURCE LEVELING AND CRITICAL RESOURCES

The critical path method of scheduling assumes unlimited resources to accomplish the project. This is true even when the schedule is resource loaded because the critical path does not take into account resource overallocation. That is, a worker can be assigned to any number of parallel activities, regardless of workload. As discussed in Best Practice 3, resource leveling adjusts the scheduled times of activities or work assignments of resources to account for the availability of resources and to improve the schedule's accuracy and credibility. Resource leveling can be relatively simple as in reassigning work from overallocated resources to underallocated resources or delaying activities until resources are available.

Leveling resources allows management to identify critical resources—those that will delay the program finish date if they are unavailable for specific activities. If resource allocation is an issue that must be addressed in the schedule, the program end date will be determined not solely through network logic but also by considering resource availability. The sequence of activities that drive the program finish date, based on both network logic and resource availability, is called the resource critical path or the resource constrained critical path or critical chain. Although the resource critical path is complex and not easily derived by even the most sophisticated schedule software, it represents the most realistic model of activities and resources that determine the minimum possible duration of the program. Once critical resources are determined, management can attempt to facilitate their work by, for example, hiring additional help, providing an uninterrupted work environment, or negotiating vacation time so that critical work is not delayed. If management focuses solely on critical activities without taking into account critical resources, it risks ignoring or overworking a program's most valuable assets or jeopardizing the project's timely completion.

CRITICAL PATH MANAGEMENT

Without clear insight into a critical path, management cannot determine which slipped activities will be detrimental to key program milestones and the program's finish date. The more complex schedules will require additional analysis by tracing critical resources. Float within the schedule can be used to mitigate critical activities by reallocating resources from activities that can safely slip to activities that must be completed on time. Until the schedule can produce a valid critical path and a valid longest path, management will not be able to provide reliable timeline estimates or identify problems or changes or their effects. Moreover, management will not be able to reliably plan and schedule the detailed work activities.

As stated earlier, the critical path and longest path must be reevaluated after each status update because the sequence of activities that make up the paths changes as activities are delayed, finish early, occur out of planned sequence, or the like. Additionally, activity duration updates and changes to logic may alter the paths. After each status update, the critical path and the longest path should be compared to the previous period's paths, and responsible resources should be alerted if previously noncritical and nondriving activities are now critical or driving. Likewise, resources that had been assigned to previously critical activities may now need to be reassigned to other critical activities. The critical and longest paths should make intuitive sense to subject matter experts. That is, the sequence, logic, and duration of critical activities should appear to be rational and consistent with the reviewers' experience.

Depending on the overall duration of the program, management may benefit by monitoring near-critical activities as well. For example, in addition to monitoring critical path activities, management may wish to monitor activities with 5 days or less total float. Monitoring near-critical activities alerts management of potential critical activities and facilitates proper resource allocation before activities become critical.[26]

Conducting a schedule risk analysis (Best Practice 8) may reveal that, with risks considered, the path most likely to delay the program is not the critical path or the longest path in the static CPM schedule. The risk analysis may identify a different path or paths that are "risk critical." Risk mitigation should focus on those risk-critical paths for best effect.

PROGRAM AND PROJECT CRITICAL PATHS

When an IMS is constructed from multiple projects, two levels of critical paths need to be managed: the overall IMS critical path and the project critical paths. When an IMS is created from several projects, the IMS-level critical path may or may not contain the same activities as those driving individual projects. The program critical path may be made of different sequences representing how each project ties into the overall program effort. However, it is expected that the program critical path will contain one project's critical path because there will be at least one project driving the overall program length.

The program critical path will change every month as work is updated and some activities are finished early and others late. Both levels of critical paths should be monitored because a failure in one project critical path can cause that project to become part of the program critical path at any time.

The management of a program-level IMS created from many projects can become daunting and highly complex, especially if software limitations prevent schedulers from easily tracking each critical path through thousands of detail activities. If the benefits of monitoring a program-level critical path through detail project activities are minimal compared to the effort required, then intermediate schedules may be used to represent some projects. As we saw in Best Practice 1, intermediate schedules are linked to lower-level detailed schedules and may or may not include detailed work activities. Projects that are an integral part of the program and contribute to the overall critical path should have their detail activities broken out. Projects that are relatively independent from the overall program duration will have their own critical paths and can be shown as summary tasks in the intermediate schedule.

---

[26]Unfortunately, no easy method of calculating the "near longest path" in complex schedules would alert management and schedulers to activities near the path of longest duration.

Best Practices Checklist: Confirming That the Critical Path Is Valid

- The schedule's critical path is valid. That is, the critical path or longest path (in the presence of constraints)

  - does not include LOE activities, summary activities, or other unusually long activities, except for future planning packages;
  - is a continuous path from the status date to the finish milestone;
  - does not include constraints that cause unimportant activities to drive a milestone date;
  - has no lags or leads;
  - is derived in summary schedules by vertical integration of lower-level detailed schedules, not by preselected activities that management has presupposed are important.

- If backward-pass date constraints are present on activities other than the finish milestone, both the critical path and the longest path have been identified. With a number of constraints, activities with zero or negative total float may outnumber activities that are actually driving the key program completion milestone.

- The critical path, or longest path (in the presence of constraints), is used as a tool for managing the program. That is, management

  - has vetted and justified the current critical path as calculated by the software;
  - uses the critical path to focus on activities that will be detrimental to the key program milestones and deliveries if they slip;
  - examines and mitigates risk in activities on the critical path that can potentially delay key program deliveries and milestones;
  - has reviewed and analyzed near-critical paths because these activities are likely to overtake the existing critical path and drive the schedule;
  - recognizes not only activities with the lowest float but also activities that are truly driving the finish date of key milestones;
  - evaluates the critical path before the schedule is baselined and after every status update to ensure that it is valid.

BEST PRACTICE 7

Ensuring Reasonable Total Float

Best Practice 7: The schedule should identify reasonable total float (or slack)—the amount of time a predecessor activity can slip before the delay affects the program's estimated finish date—so that the schedule's flexibility can be determined. The length of delay that can be accommodated without the finish date's slipping depends on the number of date constraints within the schedule and the degree of uncertainty in the duration estimates, among other factors, but the activity's total float provides a reasonable estimate of this value. As a general rule, activities along the critical path have the least float. Unreasonably high total float on an activity or path indicates that schedule logic might be missing or invalid.

Management should be aware of schedule float. Activities with the lowest total float values constitute the highest risk to completing the schedule or meeting interim milestones. In general, if zero-total-float activities or milestones are not finished when scheduled, they will delay a program the same length as their delayed finish—unless successor activities on the critical path can be completed sooner than originally planned. An activity's delay causes total float to decrease, thus increasing the risk of not completing the program as scheduled.

Incomplete, missing, or incorrect logic, unrealistic activity durations, and unstatused work distort the value of total float so that it does not accurately represent the schedule's flexibility. In addition, total float may not be a completely accurate measure of flexibility if the schedule has date constraints or deadlines such that low or even negative float values for activities do not drive the finish milestone. Thus, it is imperative that managers for both the customer and the contractor proactively manage total float as activities are completed. Doing so will ensure that the program schedule accurately depicts the program's flexibility and enables management to make appropriate decisions in reallocating resources or resequencing work before the program gets into trouble.

DEFINITIONS OF TOTAL FLOAT AND FREE FLOAT

The two types of float most commonly monitored are total float and free float.[27] Total float, the amount of time an activity can be delayed or extended before delay affects the program's finish date, can be positive, negative, or zero. If positive, it indicates the amount of time that an activity can be delayed without delaying the program's finish date.[28] Negative total float indicates the time that must be recovered so as not to delay the program's finish date beyond the constrained date. Negative total float arises when an activity's completion date is constrained—that is, when the constraint date is earlier than an activity's calculated late finish. In essence, the constraint states that an activity must finish before the date the activity may finish as calculated by network logic.

Negative float can also occur when activities are performed in a sequence that differs from the logic dictated in the network. Out-of-sequence logic is discussed in detail in Best Practice 9. Zero total float means that any amount of activity delay will delay the program's finish date by an equal amount. An activity with negative or zero total float is considered to be critical.

Free float is the portion of an activity's total float that is available before the activity's delay affects its immediate successor. Depending on the sequence of events in the network, an activity with total float may or may not have free float. For example, it may be possible that an activity slips 2 days without affecting the finish date (2 days of total float), but this delay will cause a 2-day slip in the start date of its immediate successor activity (zero free float).

Total float and free float are therefore indicators of a schedule's flexibility. Some activities in the schedule network can slip without affecting their immediate successors, and some may affect their immediate successors but not the program finish date. Knowing this allows management to reassign resources from activities that can slip to activities that cannot slip. Knowing the length of time an activity can or cannot be delayed is essential to successfully allocating resources and to completing the program on time.

Nonworking periods are not float. Nonworking periods are defined in project and resource calendars and dictate the availability of resources to work, not the flexibility of an activity's start or finish dates. In addition, float should not be treated as schedule contingency. Because float is shared along a sequence of activities, it is available for use by any activity along that sequence.

---

[27] Float, calculated from an activity's early and late dates, is the length of time the activity can be delayed before delaying the early start of its successor or the project finish date. A schedule network may also mathematically calculate *independent* float and *interfering* float. Independent float is the amount of time the activity can be delayed without delaying successor activities, given preceding activities have started late. Interfering float is the difference between an activity's total float and free float.

[28] Typically, total float is calculated by the scheduling software to the last activity in the schedule file, but other activities or interim milestones may be monitored for total float, using constraints on key interim milestones. With these constraints, using total float to identify the critical path to the finish milestone is a flawed method.

CALCULATING FLOAT

Activities on the same network path share total float. Figure 35 shows a portion of the house construction project's critical path through foundation and underground work. The critical path (in red) spans activities from "lay out and stake property and excavation" through the "pour footings and pads" activity. Two activities have total float available: "excavate for and install underground sewer" and "inspect underground sewer."

Figure 35: Total Float and Free Float

Activity	Early start	Late start	Free float	Total float	9/14/2025	9/21/2025	9/28/2025	10/5/2025
Lay out and stake property and excavation	9/15	9/15	0 days	0 days				
Dig foundation and basement	9/16	9/16	0 days	0 days				
Lay out and form footings and pier pads	9/17	9/17	0 days	0 days				
Excavate for and install underground sewer	9/16	9/22	0 days	4 days				
Inspect underground sewer	9/17	9/23	4 days	4 days				
Install footing and pier rebar	9/22	9/22	0 days	0 days				
Inspect footing and pier rebar	9/23	9/23	0 days	0 days				
Pour footings and pads	9/24	9/24	0 days	0 days				

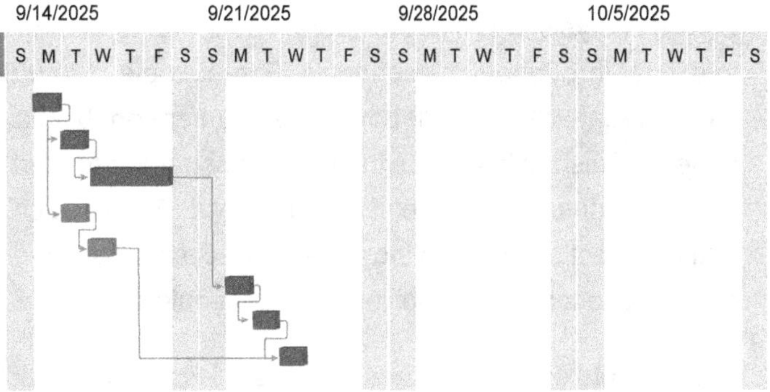

Source: GAO. | GAO-16-89G

While both "excavate for and install underground sewer" and "inspect underground sewer" have 4 days of total float available, only the inspection activity has 4 days of free float available. Any delay in excavating or installing the underground sewer will equally delay the inspection. However, inspecting the underground sewer may be delayed up to 4 days without affecting any successor activity.

Note that leveling overallocated resources on the "inspect underground sewer" activity is much easier than on "excavate and install underground sewer" because of the former activity's available free float. Delaying the inspection by 2 days affects the resources assigned to that activity, but it has no effect on any subsequent activity in the network. But delaying excavation and installation affects also inspection and resource assignments for both activities. As shown in the figure, however, leveling resources will always consume float along a path of activities.

COMMON BARRIERS TO VALID FLOAT

Unreasonable amounts of total float usually result from missing or incomplete logic rather than acceptable periods of potential delay. Therefore, any activities that appear to have a great amount of float should be examined for missing or incomplete logic. Because total float is calculated from activities' early and late dates, it is directly related to the logical sequencing of events—just like the validity of the resulting critical path. Missing activities, missing or convoluted logic, and date constraints prevent the valid calculation of total float, potentially making an activity appear as though it can slip when it actually cannot. The reasonableness of total float depends on capturing all activities (Best Practice 1) and properly sequencing activities (Best Practice 2). It also depends on realistic resource assignments (Best Practice 3) and accurate status updates (Best Practice 9). Case study 13 shows how unreasonable float values can arise from improperly sequenced activities.

Case Study 13: Unreasonable Float from the Sequencing of Activities, from *FAA Acquisitions*, GAO-12-223

GAO's analysis of FAA's Collaborative Air Traffic Management Technologies system IMS found unreasonable total float throughout the schedule. For example, 325 (68 percent) of the 481 remaining activities had float values greater than 1,000 days. These unreasonable float values were caused mostly by activities tied to the project's finish milestone, which was constrained to start no earlier than July 1, 2016. Interim milestones that were scheduled to finish earlier than July 1, 2016—such as those marking the end of task orders—were tied to the project's finish milestone as predecessors and were therefore showing enormous amounts of float that did not reflect actual flexibility in the schedule.

The majority of high-float activities were level-of-effort activities, many of which were extended to the end of these interim milestones and thus associated with unreasonable float as well. Several activities had more than 1,000 days of float, including "test and deploy," which showed a total float value of 1,280 days. This excessive float meant that delays in the activities would have no effect on the finish date of the Release 5 end milestone.

GAO, *FAA's Acquisitions: Management Challenges Associated with Program Costs and Schedules Could Hinder NextGen*, GAO-12-223 (Washington, D.C.: February 16, 2012).

Scheduling software automatically calculates total and free float for activities, which are then used to identify critical activities. However, these values of float must be examined for reasonableness. Unreasonable float might be negative, positive, or zero. That is, a network that displays large negative values of total float, such as −300 days for some activities, most likely indicates either missing logic or an unrealistic sequencing of activities. For example, dangling logic can create unrealistic free float. Because float is shared along activity paths, finding and addressing incomplete logic that causes large float values may solve the float problem for many activities on the path. Likewise, a complex schedule whose majority of remaining activities is critical is not a realistic plan and should be assessed for reasonable logic, the practicality of resource assignments, or the reasonableness of the project's duration.

REASONABLENESS OF FLOAT

Given that float is directly related to the logical sequencing of activities and indicates schedule flexibility, management and auditors will question what constitutes a reasonable amount of float for a particular schedule. Activities' float differs by status period, given the logical sequence of activities in the network and the program's remaining duration. Therefore, management should not adhere solely to a target float value (for example, maximum 2 working periods) or specific float measure (for example, 10 percent of program duration). Large amounts of float may be justified, given an activity's place in the flow of work. For example, landscaping or paving in a construction project may slip many more months than pipefitting. Likewise, nonessential activities in a 2-year project may have far more float available than the same activities in a 6-month project.[29]

In general, total float should be as accurate as possible; it should be evaluated relative to the program's projected finish date. The remaining activities in the schedule should be sorted by total float, and those values should be assessed for reasonableness. Management should determine whether it makes sense logically that any activity with relatively high float can actually slip that far without affecting the project's finish date. For instance, management should ask, is it reasonable that an activity with 55 days of total float can actually slip 55 working days before the program's finish date is affected? Is the manager of that particular activity aware of this float?

A float value of 55 days may make sense for a project that has 4 years of future planning packages, but a 55-day delay would probably be considered implausible in a 6-month project. Total float values that appear to be excessive should be documented to show that the program management team, having already performed an analysis, has agreed that the logic and float for this relevant activity are consistent with the plan. A float value that is not reasonable may result from a break in logic. Significant changes in float potentially indicate that a logic link has been broken or that an out-of-sequence activity has been completed. It may be that neither indicates true project total float.

All activities with negative float should be questioned. Negative float stems from constraining one or more activities or milestones in the network. The constraint should be examined and justified, and the resulting negative float should be evaluated for reasonableness. Management should be aware of activities that are behind schedule with respect to a constrained activity. If a delay is deemed significant, management should develop a plan to examine options for recovering from the schedule slip. If the negative float cannot be mitigated, then the milestone should forecast a slip to eliminate the negative float.

FLOAT MANAGEMENT

Total and free float calculations are fundamental products of CPM scheduling. Network logic, float, durations, and criticality of activities are interrelated. That is, the logical sequence of activities and resource assignments within a network dictate the amount of

---

[29]Total float is also tied to the overall confidence level of a schedule, assuming that a schedule risk analysis has been performed. The more optimistic a schedule may be (less confidence in meeting the completion date), the lower the available float is; the more pessimistic a schedule (more confidence in meeting the completion date), the greater the available float.

available float, and the amount of available float defines the criticality of an activity to a constrained milestone or to the final milestone, whether constrained or not. Therefore, management cannot correctly monitor the critical path without also monitoring float. Incorrect float estimates may result in an invalid critical path and an inaccurate assessment of program completion dates. In addition, inaccurate values for total float result in a false depiction of the program's true status, which could lead to decisions that jeopardize the project.

To support on-time completion of a program, management must understand the amount of time an activity can or cannot be delayed. Accurate values of total float can be used to identify activities that could be permitted to slip and resources that could be reallocated. This knowledge helps in reallocating resources optimally and in identifying the activity sequences that should be managed most closely. However, management must also balance the use of float with the fact that total float is shared along a path of activities. Allowing an activity to consume total float prevents successive activities from being able to slip and spends the schedule's flexibility rather than conserving it for future risks.

Free float is particularly important in leveling resources because leveling generally targets activities with free float first. That is, delaying an activity within its available free float will not affect its successor activity or the program's completion date. Delaying an activity that has total float but no free float does not affect the project's completion date but does disturb successor activity dates that rely on the start or finish of the delayed activity. This, in turn, may disrupt resource availability for assignments along the entire path of successor activities.

Once critical path float has been exhausted, the program is on a day-for-day schedule slip. Float on the critical path should be commensurate with program risk, urgency, technological maturity, complexity, and funding profile. Periodic reports should routinely report the amount of float consumed in a period and remaining on the critical and near-critical paths. A portent of things can be seen in the consumption of free float. As a program schedule becomes less flexible, the probability of having to consume near-critical path and critical path float is increased.

BEST PRACTICES CHECKLIST: ENSURING REASONABLE TOTAL FLOAT

- The total float values calculated by the scheduling software are reasonable and accurately reflect a schedule's flexibility.

- The program really has the amount of schedule flexibility indicated by the levels of float.

- Remaining activities in the schedule are sorted by total float and assessed for reasonableness. Any activities that appear to have a great deal of float are examined for missing or incomplete logic.

- Total float values that appear to be excessive are documented to show that the program management team has performed an assessment and agreed that the logic and float are consistent with the plan.

- Total float is calculated to the main deliveries and milestones as well as to the program's completion.

- Total and free float inform management as to which activities can be reassigned resources in order to mitigate slips in other activities.

- Management balances the use of float with the fact that total float is shared along a path of activities.

- Periodic reports routinely show the amount of float consumed in a period and remaining on the critical and near-critical paths.

- Date constraints causing negative float have been justified. If delay is significant, plans to recover the implied schedule slip have been evaluated and implemented.

BEST PRACTICE 8

Conducting a Schedule Risk Analysis

Best Practice 8: A schedule risk analysis starts with a good critical path method schedule. Data about program schedule risks are incorporated into a statistical simulation to predict the level of confidence in meeting a program's completion date; to determine the contingency, or reserve of time, needed for a level of confidence; and to identify high-priority risks. Programs should include the results of the schedule risk analysis in constructing an executable baseline schedule.

DEFINITION OF SCHEDULE RISK ANALYSIS

A schedule risk analysis uses statistical techniques to predict a level of confidence in meeting a program's completion date. This analysis focuses on uncertainty and key risks and how they affect the schedule's activity durations. Because each activity has an uncertain duration that depends in part on uncertainties about effort and resources, the entire duration of the overall program schedule is also uncertain. Therefore, unless a statistical simulation is run, calculating the completion date from schedule logic and duration estimates in the schedule tends to underestimate the overall program critical path duration.

Estimates of activity durations should be viewed as forecasts based on the best information available at the time. Assumptions regarding resource availability and productivity, required effort, and availability of materials, among other things, allow for the determination of the most likely activity durations. However, there is inherent uncertainty about the most likely duration estimate that can cause activities to shorten or lengthen. Activity duration estimates include inherent uncertainty, estimating error, and, perhaps, estimating bias. For instance, if a conservative assumption about labor productivity was used in calculating the duration of an activity and during the simulation, a better labor productivity is possible, then the activity will shorten, at least for that iteration. However, schedule underestimation is more pronounced when the schedule durations or schedule logic include optimistic bias. Activity durations and logic in a CPM schedule may be overly optimistic if there is pressure from the customer or instruction from management to finish earlier than the unbiased duration estimates imply.

Schedule Uncertainty and Risk

The terms risk and uncertainty are often used interchangeably, but they have distinct definitions in program risk analysis. *Uncertainty* refers to a situation in which little to no information is known about the outcome. A *risk* is an uncertain event that could affect the program positively or negatively. Stated another way, risk and its outcomes can be quantified in some definite way, while uncertainty cannot be defined because of ambiguity. In a situation that includes unfavorable and favorable events, the probability is that an unfavorable event will occur (a threat or harm) or that a favorable event will occur (an opportunity or improvement). Uncertainty and risk events may contain elements of both opportunity and threat.[30]

Schedule activity durations should account for both risk and uncertainty. Risk and uncertainty in scheduling refer to the fact that because activity durations are forecasts, there is always a chance that actual activity durations—and therefore scheduled start dates and finish dates—will differ from the plan. There will always be some aspect of the unknowable, and there will never be enough data available in most situations to develop a known frequency distribution of possible durations.

Risk events that can be listed and defined should be included in a program's risk register in the form of threats and opportunities. Uncertainty arises because of the inherent variability in the actions of individuals and organizations working toward a plan. Uncertainty may also include estimating error and even systematic bias, such as when estimates are consistently optimistic. These events are often called "unknown unknowns." As the program progresses, some uncertainties may be revealed or elaborated on and defined in the risk register as a threat or an opportunity. Prudent organizations recognize that uncertainties and risks can become better defined as the program advances and conduct periodic reevaluations of the risk register.

As we describe in the following sections, threats and opportunities, as well as general uncertainty, can be incorporated and quantified to some degree using schedule risk analysis.[31]

Merge Bias and Schedule Underestimation

One of the most important reasons for performing a schedule risk analysis is that the overall program schedule duration may well be greater than the sum of the path durations for lower-level activities. This is so partly because of schedule uncertainty and schedule structure. A schedule's structure has many points where parallel paths merge

---

[30]Definitions of risk and uncertainty are interrelated and vary across organizations, government agencies, and even fields of study. For example, some organizations consider risk as only the unfavorable outcome of an uncertain event.

[31]These techniques are designed to capture general uncertainties about the future, not unforeseen catastrophic events such as major earthquakes and large labor strikes.

that can cause the schedule to lengthen. Merge points may include key program events such as preliminary design review, the beginning or ending of project phases, or product deliveries. The timing of these merge points is determined by the latest merging path. Thus, if a required element is delayed, the merge event will also be delayed. Because any merging path can be risky, any merging path can determine the timing of the merge event. Figure 36 gives an example of the schedule structure that illustrates the network of a simple schedule with a merge point at start-up and test.

Figure 36: A Simple Schedule as a Network Diagram

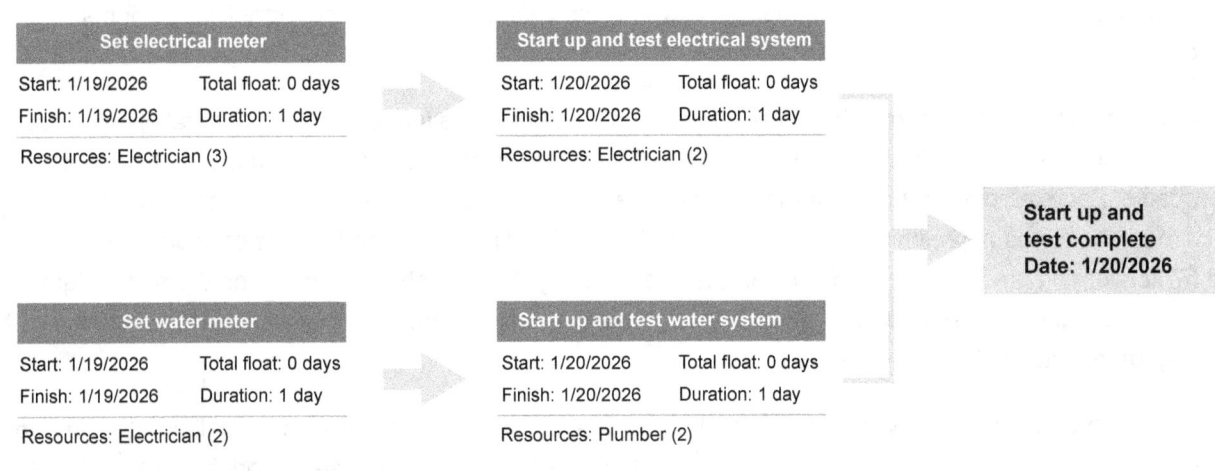

Source: GAO. | GAO-16-89G

The added risk at merge points is called "merge bias." As we discussed in Best Practice 2, risk at merge points is a concern because it is multiplicative. For example, suppose that a schedule risk analysis has determined that the two start-up and testing paths in figure 36 each has a 60 percent chance of finishing on time. The start-up paths are not necessarily a concern individually, but the success of the completion milestone is a concern. Its success is the probability of both paths completing on time—36 percent. In fact, given that each path has a 60 percent chance of success, the milestone will finish late in three of four scenarios: if the electrical test runs late but the plumbing test is on time (24 percent chance), if the electrical test is on time but the plumbing test runs late (24 percent chance), or if both the electrical and plumbing tests run late (16 percent chance).

The completion milestone is not likely to be on time even though each individual testing path is likely to complete on time. Moreover, the chance of success at a merge point decreases the more that paths converge. If a third test were added, say a furnace and air conditioning test, and its success is determined to be 60 percent also, the overall chance of success for the completion milestone would be 22 percent. Merge bias is one reason that the finish date of even a well-constructed schedule is likely to be later than scheduled. The bias is driven by a combination of risk on individual paths, the amount of free float before the milestone, and the number of merging paths at that milestone. Case study 14 provides an example of the potential effects of converging activities on scheduled activities.

Case Study 14: Converging Paths and Schedule Risk Analysis, from *Coast Guard*, GAO-11-743

The Coast Guard program office and Northrop Grumman officials said that schedule risk analysis (SRA) was not required for the National Security Cutter 3 (NSC 3) production contract and therefore it was not performed.

In the December 2010 program management review, only one risk was identified: "test or installation phase failure." Given that the schedule in February 2011 had 3,920 remaining activities, one identified risk seemed improbable. For example, Northrop Grumman officials said that the critical end milestone they were most concerned about was a "preliminary delivery of NSC." The critical milestone had 5 days of negative float and 57 converging predecessors. That is, the task was already 5 days behind schedule on the status date, and—compounding the risk of delay—had multiple converging activity paths that decreased the probability of meeting the planned milestone date.

The chance that the milestone will be accomplished on time decreases with every additional path leading up to the milestone. The more parallel paths that exist in the schedule, the greater the schedule risk is. A Monte Carlo SRA simulation could have helped identify the compound effect of parallel paths and could have quantified how much contingency reserve or margin was needed in the schedule to mitigate the risk.

Agency officials and Northrop Grumman said that a schedule risk analysis would be performed as part of the NSC 4 schedule.

GAO, *Coast Guard: Action Needed as Approved Deepwater Program Remains Unachievable*, GAO-11-743 (Washington, D.C.: July 28, 2011).

Because activity durations are uncertain, the probability distribution of the program's total duration must be determined statistically, by combining the individual probability distributions of all paths according to their risks and the logical structure of the schedule. An accepted way to do this is to perform a Monte Carlo simulation of the schedule with uncertainty and risk applied.

CONDUCTING A SCHEDULE RISK ANALYSIS

Schedule risk analysis relies on statistical simulation to randomly vary the following:

- activity durations according to their probability distribution;
- threats and opportunities according to their probability and the distribution of their effect on affected activities if they were to occur; and

- the existence of a risk or probabilistic branch.

The objective of the simulation is to develop a probability distribution of possible completion dates that reflect the program plan (represented by the schedule) and its quantified uncertainties and risks. From the cumulative probability distribution, the organization can match a date to its degree of risk tolerance.[32] For instance, an organization might want to adopt a program completion date that provides a 70 percent probability that it will finish on or before that date, leaving a 30 percent probability that it will overrun, given the schedule and the risks as they are known and calibrated. The organization can thus adopt a plan and promise completion on dates that are consistent with its preferred level of confidence in the overall integrated schedule. A schedule risk analysis can provide valuable information to senior decision makers, as shown in case study 15.

Risk analysis should not be focused solely on the deterministic critical path—that is, the critical path as defined by the initial or current set of inputs in the schedule model. Because the durations of activities are uncertain, with risk considered, any activity may potentially affect the program's completion date. Hence, the path that is most likely to determine the finish date is uncertain.

If the analysis is to be valid, the program must have a good schedule network that clearly identifies the work that is to be done and the relationships between detailed activities. The schedule should be based on a minimum number of justified date constraints. It is important to represent all work in the schedule, because any activity can become critical under some circumstances. Complete and correct schedule logic that addresses the logical relationships between predecessor and successor activities is also important. The analyst needs to be confident that the schedule will automatically calculate the correct dates and critical paths when the activity durations change, as they do thousands of times during a simulation.

If time or resources are insufficient to simulate the full program, or if detail in the future is unclear, perhaps because of rolling wave planning, the simulation can be performed with a summary version of the schedule. The summary schedule is a condensed form of the schedule that rolls detail activities into long-duration activities. By reducing the number of activities in the schedule, analysts reduce the time spent collecting data about and assigning risks and probability distributions to detail activities.

However, if a summary schedule is used for a schedule risk analysis, it is important that the schedule show enough detail to yield practical results. A summary schedule that is condensed too much will not convey the effort in very long activities, the activities

---

[32]A cumulative distribution sums all the probabilities of values less than or equal to the value of interest. The cumulative probability increases from 0 to 1 as the value of interest increases. Hence, a selected finish date from the cumulative probability distribution represents the probability of finishing on that date or earlier.

GAO performed a schedule risk analysis on the construction schedule for Phase IV of the Department of Veterans Affairs' (VA) new Medical Center Complex in Las Vegas, Nevada. The project executive identified 22 different risks in an exercise preliminary to this analysis. Using these risks as a basis for discussion, GAO interviewed 14 experts familiar with the project, including VA resident engineers, general contractor officials, and architect and engineering consultants.

In these interviews, GAO identified 11 additional risks. During data analysis, some risks were consolidated with others and some were eliminated because data were too few. Finally, 20 risks were incorporated into the Monte Carlo simulation. They included 18 risk drivers, 1 schedule duration risk, and 1 overall system commissioning activity that was not included in the baseline schedule.

The schedule duration risk was applied to each activity duration to represent the inherent variability of project activities and inaccuracy of scheduling. Of the 6,098 activities in the schedule, GAO assigned risk drivers to 3,193. Some activities had one or two risks assigned, but some had as many as seven.

Beyond applying 20 risks to the schedule, GAO was interested in identifying the marginal effect of each risk. Therefore, GAO identified the risks that had the greatest effect on the schedule, because they should have been targeted first for mitigation. Marginal effect translates directly to potential calendar days saved if the risk is mitigated.

GAO's analysis of the medical center construction schedule concluded that VA should have realistically expected VA's acceptance between March 1, 2012, and May 17, 2012, the 50th and 80th percentiles. It was determined that the must-finish date of August 29, 2011, was very unlikely. The analysis showed that the probability of achieving VA's acceptance by October 20, 2011, was less than 1 percent, given the current schedule without risk mitigation.

VA's actual acceptance was December 14, 2011, approximately 4 months later than had originally been expected. Delays stemmed from issues with steel fabrication and erection, as well as changes to equipment requirements. At the time of GAO's original analysis, December 14, 2011, fell within the 5th to 10th percentiles.

GAO, *VA Construction: VA Is Working to Improve Initial Project Cost Estimates, but Should Analyze Cost and Schedule Risks*, GAO-10-189 (Washington, D.C.: December 14, 2009).

that should have assigned risks, or how total float is distributed among key activities and milestones. For example, activities in the summary version of the schedule should show critical hand-offs. If an activity is 4 months long but a critical hand-off is expected halfway through, the activity should be broken down into separate 2-month activities that logically link the hand-off between activities. Finally, condensing the schedule may hide merging paths. As discussed in the previous section, merging paths are the source of much risk.

After the risk information is developed, the statistical simulation is run and the resulting cumulative distribution curve, the S curve, displays the probability associated with the range of program completion dates. The results of risk analysis are best viewed as inputs to program management rather than as forecasts of how the program will be completed. The results indicate when the program is likely to finish without the program team's taking additional risk mitigation steps. The high-priority risks can be identified and used to guide further risk mitigation action.

A schedule risk analysis may show that a given schedule has more risk than is acceptable. In this case, a review of the activities, dependencies, and network might help derive a shorter schedule. In some cases, the scope may need to be reduced. However, the initial estimates of effort and duration should not be changed without sufficient justification. Changing durations simply because an earlier finish date is preferred is likely to increase the risk of delaying a project.

COLLECTING ANONYMOUS AND UNBIASED RISK DATA

A schedule risk analysis requires the collection of program risk data. Risk data should be derived from a quantitative risk assessment and should not be based on arbitrary percentages or factors. A risk assessment is a part of the program's overall risk management process in which risks are identified and analyzed and the program's risk exposure is determined. As risks are identified, risk-handling plans are developed and incorporated into the program's cost estimate and schedule, as necessary.[33]

Risk data can be collected in the form of three-point durations or by using the risk driver approach, to be described in the next section. The three-point estimates represent inherent uncertainty, estimating error, and perhaps estimating bias, while risk drivers represent identified risk events with probabilities as well as the likely effect if they occur. Regardless of which type of risk analysis is performed, it is essential that subject matter experts (SME) be interviewed who are directly responsible for or involved in the workflow activities. Estimates derived from interviews should be formulated in a consensus of knowledgeable technical experts and should be coordinated with the same people who manage the program and its risk mitigation watch list. Employees involved in the program from across the entire organization should be considered for interviewing. Lower-level employees have valuable information on day-to-day tasks in specific areas of the program, including their insight into how individual risks might affect their workflow responsibilities. Managers and senior decision makers have insight into all or many areas of the program and can provide a sense of how risks might affect the program as a whole.

The starting point for the risk interviews is the program's existing risk register. Interviewees are asked to provide their opinion on threats and opportunities and should be en-

---

[33]For more information on formal risk assessments and their relation to cost, schedule, and EVM, see GAO-09-3SP.

couraged to introduce additional risk events that are not on the risk register. If unbiased data are to be collected, interviewees must be assured that their opinions on threats and opportunities will remain anonymous. They should also be guaranteed nonattribution and should be provided with an environment in which they are free to brainstorm on worst and best case scenarios. It is particularly important to interview SMEs without an authoritative figure in the room to avoid motivational bias.

Motivational bias is a source of bias that arises when interviewees feel threatened (whether justifiably or nonjustifiably) if they give their true thoughts about a program. This threat is typically from fear of being punished by someone in authority. Most commonly, interviewees are labeled trouble makers or are ostracized from the team if their worst case scenario is worse than management's opinion. Risk workshops may exhibit social and institutional pressures to conform, perhaps to get consensus or to shorten the interview session. The organization may greatly discourage introducing a risk that has not been previously considered, particularly if the risk is sensitive or may negatively affect the program. If an interviewee is accompanied by someone, risk analysts cannot guarantee that the interviewee's responses are unbiased.

SCHEDULE RISK ANALYSIS WITH THREE-POINT DURATION ESTIMATES

One way to capture schedule activity duration uncertainty is to collect various estimates from individuals and, perhaps, from a review of actual program performance. Table 2 shows a traditional approach with a three-point estimate applied directly to the activity durations for a section of the house construction schedule. The example shows three-point estimates of remaining durations. In an actual program schedule risk analysis, these would be developed from in-depth interviews of persons who are knowledgeable in each of the WBS areas of the program.

Table 2: Estimated Durations for a Section of the House Schedule

ID	Description	Minimum remaining duration	Most likely remaining duration	Maximum remaining duration
A1870	Install drywall on walls and ceilings	3	4	6
A1880	Inspect drywall screws	1	1	2
A1890	Finish drywall (tape and mud)	3	5	6
A1900	Install ceiling insulation	1	1	2
A1910	Apply drywall texture	2	3	4
A1920	Apply wall finishes (stain and paint)	2	3	4
A1930	Install tile in bathroom and kitchen	2	3	5

Source: GAO | GAO-16-89G.

To model the risks in the simulation, the risks are represented as triangular distributions specified by the three-point estimates of the activity durations. In other words, for this example the 3-point estimates represent all the risk in the construction project. Distributions other than the triangular are traditionally available as well.[34]

Once the distributions have been established, a statistical simulation (typically a Monte Carlo simulation) uses random numbers to select specific durations from each activity probability distribution, and a new critical path and dates are calculated, including major milestone and program completion dates. The Monte Carlo simulation continues this random selection thousands of times, creating a new program duration estimate and critical path each time. The resulting frequency distribution displays the range of program completion dates along with the probabilities that activities will occur on these dates, as seen in figure 37.

Figure 37: The Cumulative Distribution of the House Construction Schedule

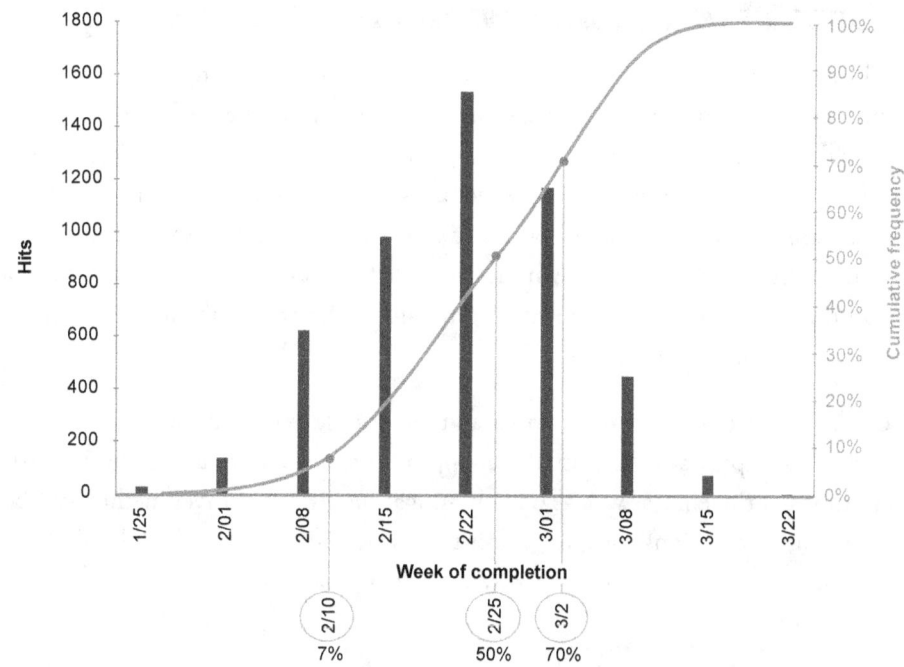

Source: GAO. | GAO-16-89G

The figure shows that the expected completion date is February 25, not February 10, which is the date the deterministic schedule computed. The cumulative distribution shows that, in this instance, the likelihood is about 7 percent that the project will finish on February 10 or earlier, given the schedule and the risk ranges used for the durations. Moreover, a contractor planning for 70 percent certainty would promise completion on March 2, about a calendar month later than originally planned.

---

[34]For more information on developing probability distributions to model uncertainty, see chapter 14 in the GAO *Cost Estimating and Assessment Guide*, GAO-09-3SP.

Three-point duration risk analyses, an acceptable method of conducting SRAs, are widely used. However, a disadvantage of using three-point duration ranges to represent all the risk in an analysis is that probability distributions for durations derived from risk interviews cannot be attributed to individual risk events. Interviewees may be combining any number of threats and opportunities in their single best case and worst case estimates. For example, a construction manager may suggest a worst-case scenario of 6 days to install drywall, as shown in table 2. However, the delay may be caused by lack of materials, poor labor productivity, poor weather, last-minute design changes, or some serial combination of all four risks. It is also possible that the SME has increased the pessimistic duration estimate to account for general uncertainty, in effect accounting for "unknown unknowns." The result of the three-point duration SRA is a recommended amount of schedule contingency that covers both quantified risks and some level of uncertainty but gives no indication of which risks are most likely to affect the schedule.

SCHEDULE RISK ANALYSIS WITH RISK DRIVERS

A second way to determine schedule activity duration uncertainty is to analyze the probability that risks from the risk register may occur and what their effect on schedule activities will be if they do occur. With this approach, a probability distribution of the risk impact—expressed as a multiplicative factor—on the duration of activities in the schedule is estimated and the risks are assigned to specific activities in the schedule. If a risk does not occur in an iteration, then the scheduled duration does not change for that activity. In this way, activity duration risk is estimated indirectly by the root cause risks and their assignments to activities.

A risk can be assigned to multiple activities and the durations of some activities can be influenced by multiple risks. This risk driver approach focuses on risks and their contribution to time contingency as well as on risk mitigation. The risk driver method can be used to examine how various risks may affect the house construction schedule. Table 3 shows a subset of possible risks associated with the construction.

Table 3: Some Identified Risks for a House Construction Schedule

Risk	Likelihood of risk	Effect on remaining duration		
		Optimistic	Most likely	Pessimistic
Design is incomplete	80%	95%	125%	150%
Site investigation is inadequate	30	100	120	135
Material is unavailable	25	100	125	130
Material is late or defective	35	95	110	130
Inspectors are unavailable	30	125	150	200
Rework will be necessary	25	100	110	135
Materials are purchased incorrectly	10	100	110	130
Soil conditions are poor	25	100	115	135
Owner makes changes	50	95	110	130

Source: GAO | GAO-16-89G.

According to table 3, we can suspect that the biggest risk in the construction schedule involves design and that the plan may be too aggressive in assuming that the design will be completed early. Moreover, late or defective materials and changes by the owner are also likely to affect the schedule.

In addition to including discrete threats and opportunities, we can include risks that represent ambiguity about the future. The existence of these ambiguities is known (their likelihood is 100 percent) but their effects are unknown. For example, we know that the productivity of labor will affect the duration of many activities, but whether the overall effect is positive (an opportunity) or negative (a threat) is unknown. We can also include some element of general uncertainty. For example, we know that natural variability surrounds each of our duration estimates, so we include an uncertainty to represent a global estimating error. Table 4 identifies some uncertainties for the house construction schedule.

Table 4: Some Uncertainties for a House Construction Schedule

Uncertainty	Likelihood of risk	Effect on duration		
		Optimistic	Most likely	Pessimistic
Productivity of labor	100%	95%	100%	110%
Efficacy of general contractor	100	90	100	125
Schedule estimating error	100	95	105	115

Source: GAO | GAO-16-89G.

With the risk driver method, the risks shown in tables 3 and 4 will appear as factors that multiply the durations of the activities they are assigned to, if they occur in the iteration. Once the risks are assigned to activities, a simulation is run. The results may be similar to those in figure 38.

Figure 38: House Construction Schedule Results from a Risk Driver Simulation

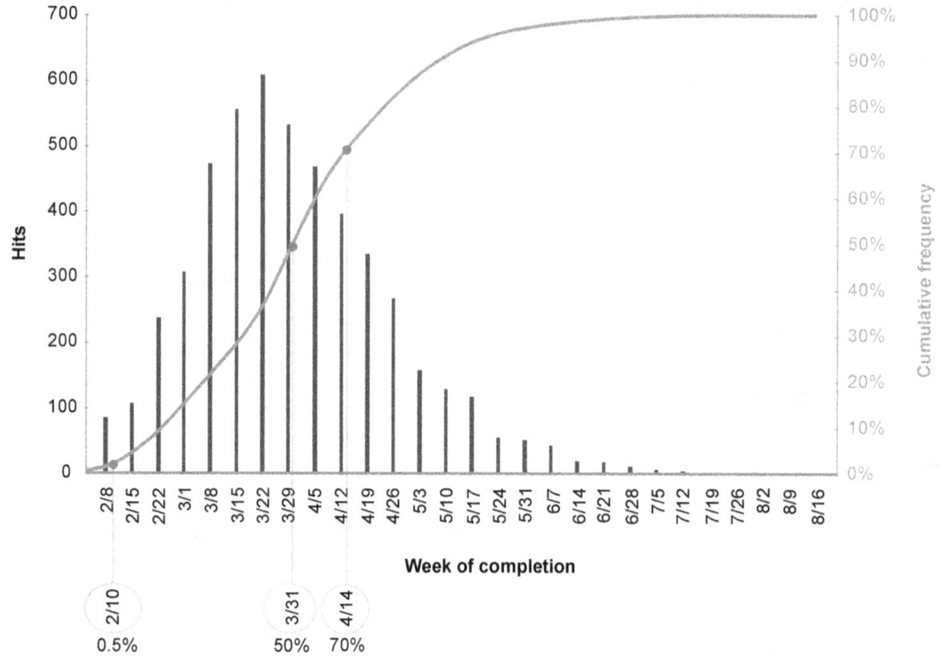

Source: GAO. | GAO-16-89G

In this instance, the schedule date of February 10 is estimated to be less than 1 percent likely, based on the current plan. If the owner chose the 70th percentile, the date would be April 14, representing a 2-month time contingency. Notice that the risk driver method has caused a wider spread of uncertainty between the 5 percent and 95 percent confidence dates compared to the three-point duration method. By combining the two methods, three-point estimates may be used to represent bias and uncertainty, while risk drivers are used to represent identifiable risk events that may be mitigated.

PRIORITIZING RISKS

No program can mitigate all risk and uncertainty. Some risks may be highly probable yet cause a relatively small delay to the finish date. Conversely, a risk may potentially delay the program a long time but be highly unlikely to ever occur. In addition, it is impossible to fully mitigate uncertainty because of its inherent ambiguity. Therefore, regardless of the method used to examine schedule activity duration uncertainty, it is important to identify the risks that contribute most to the program schedule risk. These risks can then be targeted for mitigation strategies.

Sensitivity measures reflecting the correlation of the activities or the risks with the final schedule duration can be produced by most schedule risk software. Figure 39 shows a standard schedule sensitivity index for the house construction project.

Figure 39: Sensitivity Indexes for the House Construction Schedule

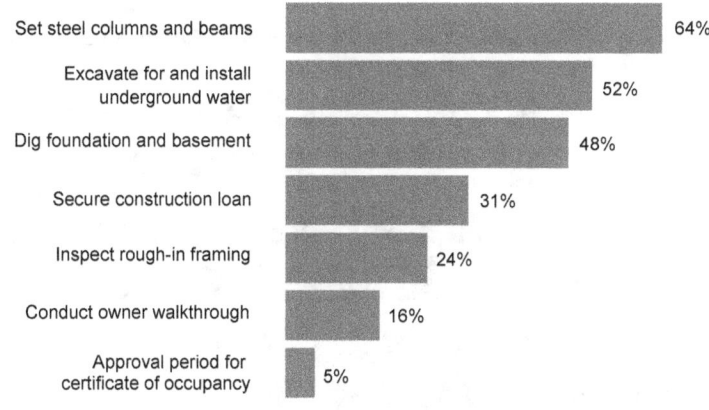

In the example in figure 39, setting the steel columns and beams affects the schedule duration more than digging the foundation, securing the construction loan, or the length of the approval period for the certificate of occupancy. The duration sensitivity chart identifies activities and paths that tend to be associated with project risk.

Figure 40 is a risk tornado chart, showing the correlation between a risk driver and project duration. It shows that when a risk is assigned to several activities, its sensitivity measure reflects the entire correlation, not just the correlation of one activity to the project's duration. According to this analysis, incomplete design is the biggest risk driver in the house construction schedule, followed by the availability of materials. Using this information, the owner and general contractor can work together to ensure that the design is complete before the project begins, as well as identify alternative sources for key materials.

Figure 40: Evaluation of Risk Sensitivity in the House Construction Schedule

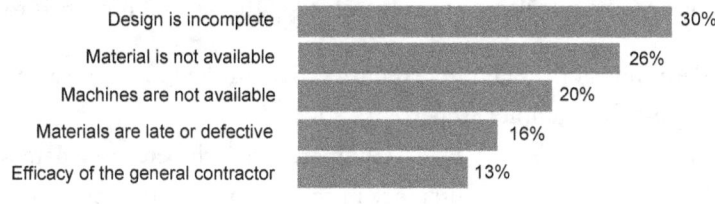

Risk analysis should also identify the activities that most often ended up on the critical path during the simulation, so that near risk-critical path activities can be identified and closely monitored. Risk criticality represents the percentage of simulation iterations that an activity or milestone is on the critical path. Figure 41 shows risk criticality for selected activities in the house construction schedule.

Figure 41: Risk Criticality of Selected Activities in the House Construction Schedule

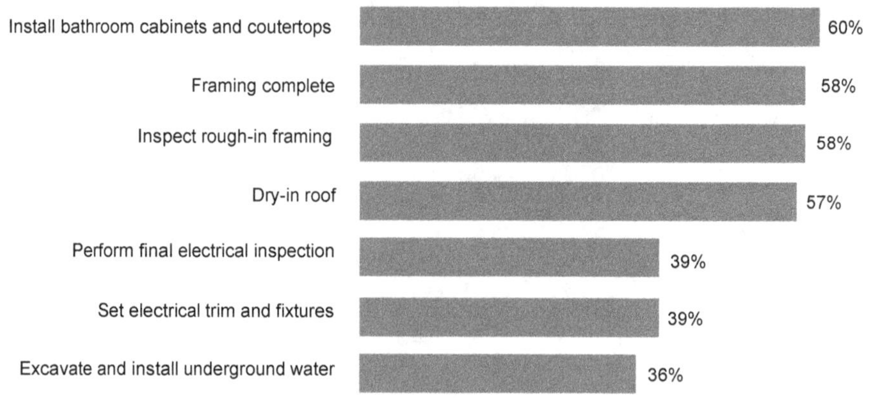

In figure 41, the activities most likely to be on the critical path may not be the most risky themselves. The activities may be critical because they are appearing on a path whose criticality is driven by some risk affecting other activities.

Sensitivity indexes and correlation measures are useful starting points for assessing the possible magnitude of realized risks, but they have limited use in prioritizing risks. Notice that figure 39 shows activities, not the root cause risks. Thus, while the chart is useful for indicating where risk is the greatest, it cannot be used to identify specific risks for mitigation. And while figure 40 gives the correlations between risk drivers and project duration, no measure of time or cost is associated with the risks.

If the risk drivers method is used, the risks can be prioritized by their effect on the risk of finishing on time and their share of required contingency. If one risk at a time is removed and the Monte Carlo simulation is rerun, the contribution of each risk to the required contingency can be calculated at any percentile. The general process is to

- Run the SRA using all risks and uncertainties. Record the finish date at the desired percentile—for example, 80 percent.
- Remove a risk and run the SRA again. Compare the 80th percentile date with the date of the full model. The difference in the two dates is the expected contribution (or days saved) of the removed risk.
- Replace the risk, remove the next risk, rerun to the SRA, and again compare the 80th percentile date to the date from the full SRA simulation and calculate the difference in days. Continue removing risks and uncertainties one by one; the risk with the highest contribution in days is the most important risk.

To identify the next most important risk, the most important risk identified above is removed from the model and the process is repeated. As the risks are removed and the next most important risks are identified, a prioritized list of risks and uncertainty can be created, similar to table 5. Table 5 shows the top 5 risks and their contribution to the 80th percentile date of April 23, 2026.

Table 5: Top Prioritized Risks in the House Construction Schedule

Priority	Risk	80th percentile date	Calendar days saved
	All	April 23, 2026	
1.	Design is incomplete	April 1	22
2.	Machines are not available	March 23	9
3.	Material is not available	March 17	6
4.	Efficacy of the general contractor	March 10	7
5.	Materials are late or defective	March 4	6

Source: GAO | GAO-16-89G.

The results in table 5 are the same as the risks identified in figure 40, although not in the same order. The order is likely to be different because as risks are removed from the simulation, activity durations and critical sequences change, depending on the nature of the schedule network and how risks and uncertainty were assigned to activities.

PROBABILISTIC BRANCHING

In addition to standard schedule risk and sensitivity analysis, typical events in programs require adding some new activities to the schedule. This is called "probabilistic branching." One common event is the completion of a test of an integrated product (for example, a software program or satellite). A schedule often assumes that tests are successful, whereas experience indicates that tests may fail and that their failure will require the activities of root cause analysis, plan for recovery, execution of recovery, and retest. This is a branch that happens only with some probability.[35]

In the house construction example, the SRA accounts for two scenarios that could occur after owner walkthrough. The plan assumes that in 70 percent of the cases, deficiencies identified during walkthrough can be addressed by the general contractor within a work week. However, in 30 percent of the cases, the owner and the general contractor dispute the deficiencies for 15 working days, and, unable to resolve their differences, enter mediation for 30 working days. The Gantt chart in figure 42 shows the probabilistic branch associated with owner walkthrough for this example.

---

[35] Probabilistic branching analyses may need to be conducted on a copy of the IMS file if the IMS is baselined or represents only required scope.

Figure 42: Probabilistic Branching in a Schedule

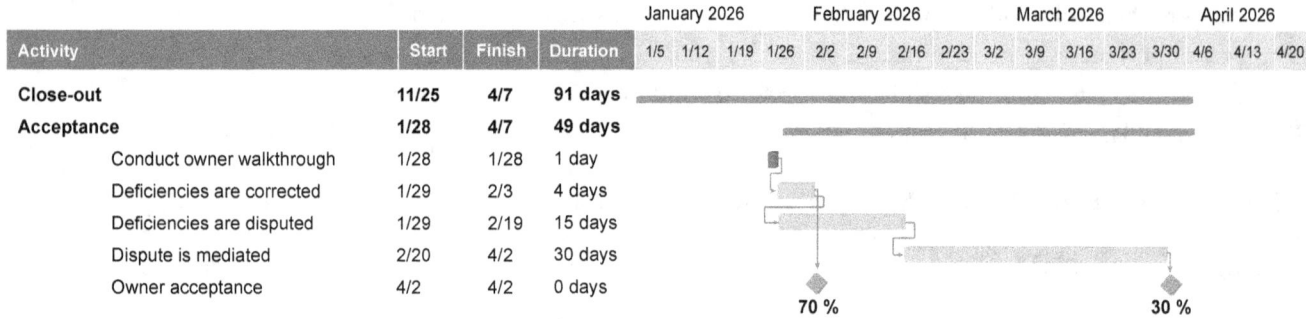

Activity	Start	Finish	Duration	January 2026	February 2026	March 2026	April 2026
Close-out	**11/25**	**4/7**	**91 days**				
Acceptance	**1/28**	**4/7**	**49 days**				
Conduct owner walkthrough	1/28	1/28	1 day				
Deficiencies are corrected	1/29	2/3	4 days				
Deficiencies are disputed	1/29	2/19	15 days				
Dispute is mediated	2/20	4/2	30 days				
Owner acceptance	4/2	4/2	0 days				

Source: GAO. | GAO-16-89G

In figure 42, "deficiencies are corrected" will occur 70 percent of the time, resulting in no delay for owner acceptance. In 30 percent of cases, "deficiencies are disputed" occurs, leading to the successor activity "dispute is mediated." This results in a delay of 40 working days to owner acceptance and, ultimately, in a 40-working day delay to owner occupation. The resulting probability distribution of dates for the entire project can be depicted as in figure 43, as applied to the 3-point duration risk simulation.

Notice the bimodal distribution with the corrected deficiencies scenario on the left of figure 43 and the dispute scenario on the right. In this case, if the homeowner demanded an 80th percentile schedule, it would be April 15.[36]

---

[36]Probabilistic branching is used to model the random choice between two alternatives. An advanced technique known as "conditional branching" is also available in certain SRA software packages. With conditional branching, an action is determined by some scheduled event rather than by randomness. That is, it is modeled as an "If…then…else" statement rather than a probability of occurrence. For example, if a design activity takes 2 weeks, then execute Plan A, otherwise (else) execute Plan B.

Best Practice 8: Conducting a Schedule Risk Analysis

Figure 43: Probability Distribution Results for Probabilistic Branching

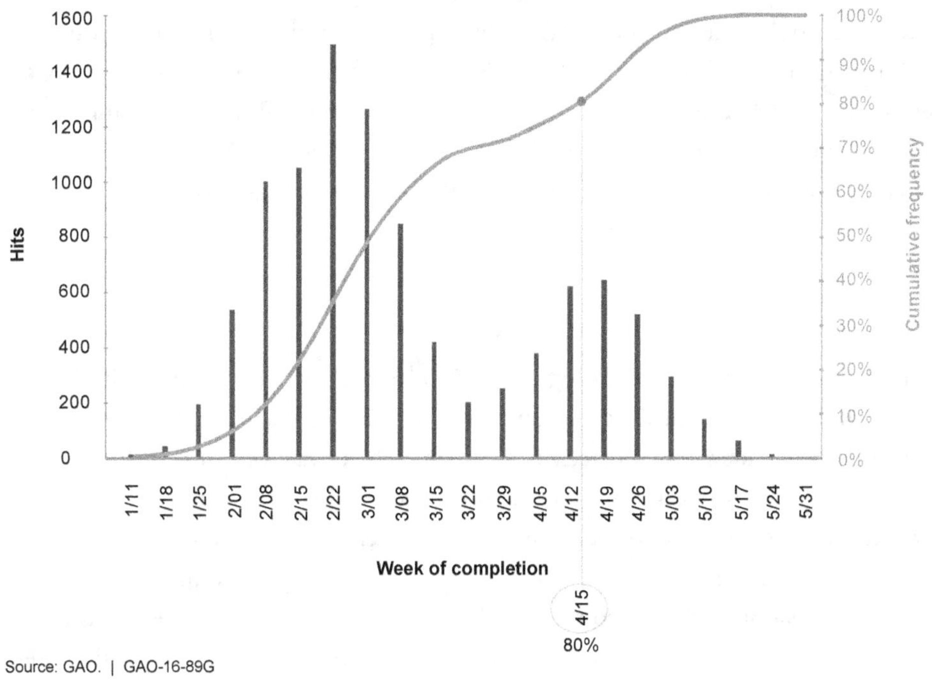

Source: GAO. | GAO-16-89G

CORRELATION

Other capabilities are possible once the schedule is viewed as a probabilistic statement of how the program might unfold. One that is notable is the correlation between activity durations. Positive correlation is when two activity durations are both influenced by the same external force and can be expected to vary in the same direction within their own probability distributions in any consistent scenario.[37] Correlation might be positive and fairly strong if, for instance, the same assumption about the maturity of a technology is made to estimate the duration of design, fabrication, and testing activities or the contractor's productivity affecting multiple activities that have been bid. If the technology maturity is not known with certainty, it would be consistent to assume that design, fabrication, and testing activities would all be longer or shorter together.

Likewise, if a particular trade is relatively unproductive in the house construction example, we may expect all activities associated with that trade to be delayed to some degree. Without specifying correlation between these activity durations in simulation, some iterations or scenarios would have some activities that are thought to be correlated go long and others short in their respective ranges during an iteration. This would be inconsistent with the idea that they all react to the same assumptions about technology maturity or trade productivity.

---

[37]While durations might vary in opposite directions if they are negatively correlated, this is less common than positive correlation in program management.

Specifying correlations between related activities ensures that each iteration represents a scenario in which their durations are consistently long or short in their ranges together. Because schedules tend to add durations (given their logical structure), if the durations are long together or short together, there is a chance that projects will be very long or very short. Correlation affects the low and high values in the simulation results. This means that the high values are even higher with correlation and the low values are even lower, because correlated durations tend to reinforce one another down the schedule paths. In practice, if the organization wants to focus on the 80th percentile, correlation matters; correlation does not matter as much around the mean duration from the simulation.

Figure 44 shows the effect of adding correlation between activity durations in the three-point risk simulation for the house construction schedule. In this example, 90 percent correlation was added between activities that are related trades. While the 90 percent correlation is high (correlation is measured between –1.0 and 1.0), there are often no actual data on correlation, so expert judgment is often used to set the correlation coefficients. Assuming this degree of correlation, we get the result shown in figure 44. Notice that the correlation has widened the overall distribution. The 50th percentile is nearly the same in both cases, February 25 without correlation and February 24 with correlation. However, the 80th percentile increases by one workweek, from March 4 to March 9, when correlation is added.

Figure 44: Probability Distribution Results for Risk Analysis with and without Correlation

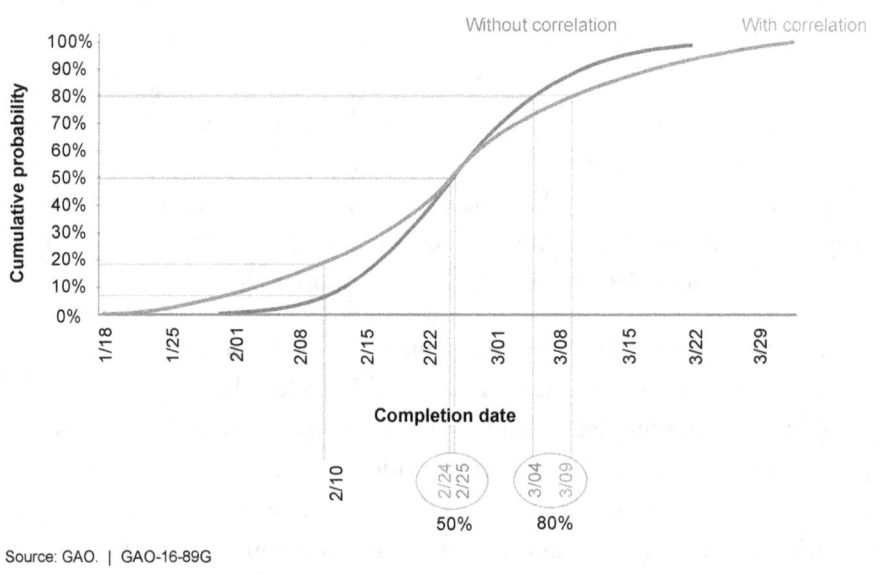

Source: GAO. | GAO-16-89G

Using three-point estimates for activity durations requires estimating correlation coefficients, often in the absence of historical data. Inconsistent correlation matrixes

Best Practice 8: Conducting a Schedule Risk Analysis

often result in this pair-wise setting of correlation coefficients. In the risk driver method, assigning a risk to multiple activities causes them to be correlated, because if the risk occurs on one assigned activity during the simulation, it occurs on all the assigned activities. If there are also some risks on one activity but not another, correlation will be less than 100 percent. Modeling correlation with risk drivers avoids the difficult task of estimating a number of pair-wise correlations.

SCHEDULE CONTINGENCY

A baseline schedule includes margin or a reserve of extra time, referred to as schedule contingency, to account for known and quantified risks and uncertainty. The contingency represents a gap in time between the finish date of the last activity (the planned date) and the finish milestone (the committed date). When schedule contingency is depicted this way, a delay in the finish date of the predecessor activity results in a reduction of the contingency activity's duration. This reduction translates into the consumption of schedule contingency.

Schedule contingency should be calculated by performing a schedule risk analysis and comparing the schedule date with that of the simulation result at a desired level of certainty. For example, an organization may want to adopt an 80 percent chance that its program will finish on time or earlier. The amount of contingency necessary would be the difference in time between the 80th percentile date from the cumulative distribution and the date of the deterministic finish date in the schedule.

For some programs, the 80th percentile is considered a conservative promise date. Other organizations may focus on another probability, such as the 65th or 55th percentile. However, because schedule distributions tend to be right skewed (that is, the program has a greater tendency to finish late than early), the mean of the distribution tends to be larger than the 50 percent confidence level. Hence, the 55th or 65th percentiles are not as certain as the 80th percentile and may expose the program to overruns if they are adopted. Factors such as project type, contract type, and technological maturity affect each organization's determination of its tolerance for schedule risk.

Schedule contingency or reserve is held by the program manager but can be allocated to contractors, subcontractors, partners, and others as necessary for their scope of work. When contingency needs to be allocated, a formal change process should be followed. Subjecting schedule contingency to the program's change control process ensures that variances can be tracked and monitored and that the use of contingency is transparent and traceable.

Schedule contingency may appear as a single activity just before the finish milestone or it may be dispersed throughout the schedule as multiple activities before key milestones. For example, it might be appropriate to plan a contingency activity before the start of a key integration activity that depends on several external inputs to ensure readiness to start. It is preferable that contingency be held as one activity just before the finish milestone for several reasons. In general, apportioning contingency in advance to specific activities is not recommended because risks that will actually occur and the magnitude of their effects are not known in advance. In addition, dispersing contingency to specific key milestones may cause its consumption prematurely or superfluously.

Contingency that is dispersed throughout the schedule is less visible and may be harder to track and monitor. Dispersion may also encourage team members to work toward late dates rather than the expected early dates. By aggregating contingency, everyone on the project will be working to protect the schedule contingency as a whole, not simply their own portions. Finally, if contingency activities are dispersed within the schedule network, care must be taken that the contingency activities do not affect total float and, therefore, critical path calculations. Contingency activities should not become critical because no resources or scope are associated with them and they cannot practically delay a successor activity.[38]

Regardless of whether contingency is captured at the end of the schedule or just before key milestones, representing it as activities will help ensure that the schedule is not hiding potential problems. Contingency can also be quickly identified and zeroed out before schedule health measures are calculated or an SRA is conducted if it is represented as activities.[39] Finally, schedule contingency should not be represented as a lag between two activities. Lags have no descriptive name in schedules and the associated contingency may become lost within the network logic.

Notice that contingency is not the same as total float. As described in Best Practice 7, total float is the amount of time by which an activity can be delayed before it affects the finish milestone. Total float is directly related to network logic and is calculated from early and late activity dates. Schedule contingency, in contrast, is determined by a schedule risk analysis. A schedule risk analysis compares the schedule date with that of the simulation result at a desired level of certainty and is calculated by quantifying uncertainties and risks that may affect the finish date.

UPDATING AND DOCUMENTING A SCHEDULE RISK ANALYSIS

A schedule risk analysis should be performed on the schedule periodically as the schedule is updated to reflect progress on activity durations and sequences. As the program progresses, risks retire or change in potential severity, and new risks that were previously

---

[38]A technique for inserting contingency activities within the network is to insert an activity as a predecessor to a "reference milestone." The reference milestone is a copy of the original key milestone but has no successor. This method allows the reference milestone to shift in response to the depletion or accumulation of contingency without affecting total float values in the schedule.

[39]If activities representing schedule contingency are not removed before an update to the SRA is conducted, they will adversely affect the results.

categorized as "unknown unknowns" may appear. The time between SRA updates will vary according to program length, complexity, risk, and the availability of management resources. A contractor should perform an SRA during the formulation of the performance measurement baseline to provide the basis for contractor schedule reserve at the desired confidence level. Preferably, an SRA is also performed before key decision points throughout the program. An SRA might occur more regularly, for instance, to support annual budget request submissions so that adequate contingency can be included in the budget baseline.

SRAs should also be performed as needed, as when schedule challenges begin to emerge with a contractor and when schedule contingency is consumed at a higher-than-expected rate or is consumed by the materialization of risks not included in the risk register. The risk register should be updated with any new risks identified during risk analysis data collection. An updated SRA is particularly important to support the internal independent assessment process if the program is rebaselined or if significant changes are made to the risk register. Keeping the program schedule current is discussed in Best Practice 9 and rebaselining is discussed in Best Practice 10.

Each update to the SRA should be fully documented to include the risk data, sources of risk data, and techniques used to validate the risk data. In addition, the methodologies used to perform the simulation should be detailed, and outputs such as a prioritized risk list, the likelihood of the program completion date, the activities that most often ended up on the critical path, and the derivation of contingency sufficient for risk mitigation should be documented.

BEST PRACTICES CHECKLIST: CONDUCTING A SCHEDULE RISK ANALYSIS

- A schedule risk analysis was conducted to determine

  - the likelihood that the program completion date will occur,
  - how much schedule risk contingency is needed to provide the acceptable certainty of completion by a specific date,
  - risks most likely to delay the project, and
  - the paths or activities that are most likely to delay the program.

- The schedule was assessed against best practices before the simulation was conducted. The schedule network clearly identifies work to be done and the relationships between detailed activities and includes a minimum number of justified date constraints.

- The SRA has optimistic, most likely, and pessimistic duration data fields.

- The SRA accounts for correlation in the uncertainty of activity durations.

- Risks are prioritized by probability and magnitude of effect.

- The risk register was used in identifying the discrete risks potentially driving the schedule before the SRA was conducted.

- The SRA data and methodology are available and documented.

- The SRA identifies the activities in the simulation that most often ended up on the critical path, so that near-critical path activities can be closely monitored.

- The risk inputs have been validated. The probabilities and impact ranges are reasonable and based on information gathered from knowledgeable sources, and there is no evidence of bias in the risk data.

- The baseline schedule includes schedule contingency to account for the occurrence of risks. Schedule contingency is calculated by performing an SRA and comparing the schedule date with that of the simulation result at a preferred level of certainty.

- Schedule contingency is held by the program manager and allocated to contractors, subcontractors, partners, and others as necessary for their scope of work.

- The program documents the derivation and amount of contingency management has set aside for risk mitigation and unforeseen problems. An assessment of schedule risk is performed to determine whether the contingency is sufficient.

- A contractor performs an SRA during the formulation of the performance measurement baseline to provide the basis for contractor schedule reserve at the preferred confidence level.

- An SRA is performed on the schedule periodically as the schedule is updated to reflect actual progress on activity durations and sequences, as well as new risks.

BEST PRACTICE 9

Updating the Schedule Using Actual Progress and Logic

Best Practice 9: Progress updates and logic provide a realistic forecast of start and completion dates for program activities. Maintaining the integrity of the schedule logic is necessary to reflect the true status of the program. To ensure that the schedule is properly updated, people responsible for the updating should be trained in critical path method scheduling.

"Statusing" is the process of updating a plan with actual dates, logic, and progress and adjusting forecasts of the remaining effort. Statusing the schedule is fundamental to efficient resource management and requires an established process to provide continual and realistic updates to the schedule. Updates should be regular and fully supported by team members and program management.

The benefits of updating the schedule on a regular basis include

- knowledge of whether activities are complete, in progress, or late and the effect of variances on remaining effort;
- continually refined duration estimates for remaining activities using actual progress, duration, and resource use;
- the current status of total float and critical path activities; and
- the creation of trend reports and analyses to highlight actual and potential problems.

The time between status updates depends on the program's duration, complexity, and risk, as well as the detail in the IMS. Updating the schedule too often will misuse team members' time and will provide little value to management, but updating too infrequently makes it difficult to respond to actual events. The schedule may be updated less frequently in the beginning of a program, then more frequently as it progresses and as more resources begin working on activities.

Program managers should consider tracking progress in the schedule more frequently than the reporting period. For example, if the reporting period is monthly, schedule progress should be updated weekly. If schedule status is being reported weekly, then progress should be tracked daily. Tracking progress at a lower level than the reporting period allows management insight into the causes of issues that are being reported before the end of the reporting period. In Best Practice 4, we discuss the importance of keeping near-term durations shorter than the reporting period.

The schedule should reflect actual progress as well as other information such as actual start and finish dates, forecasted dates, and logic changes. Activity owners provide these data to the scheduler. To ensure that the schedule is properly updated, responsibility for changing or statusing the schedule should be assigned to someone who has the proper training and experience in CPM scheduling. Certain scheduling software packages may appear to be easy to use at first, but a schedule constructed by an inexperienced user may hide or ignore fundamental network logic errors and erroneous statusing assumptions. Once an update has been made, management should assess its accuracy to verify that all finished work is in the past and all unfinished work is scheduled for the future.

Statusing the schedule should not be confused with revising the schedule. Statusing involves updating the schedule with actual facts and comparing those facts against a plan, such as events on certain dates and the progress of work with a certain number of resources. After the schedule is statused, management may want to revise the plan for remaining work by, for example, changing the sequence of activities or adding activities. That is, statusing the schedule involves updating it with actual data, while revising the schedule focuses on adjusting future work.

Appropriate time for revising the schedule must be included in the update process. Otherwise, maintaining the stability of the schedule will be difficult. It is useful for teams to create calendars for scheduling updates and revisions. The concepts of altering the plan based on knowledge gained or actual performance are referred to as either replanning or rebaselining, depending on the program's approach to change control. These concepts are discussed further in Best Practice 10.

STATUSING PROGRESS

Statusing progress generally takes the form of updating to either durations or work and includes updating the status of remaining work and network logic. A status date (or data date) denotes the date of the latest update to the schedule and thus defines the demarcation between actual work performed and remaining work. All work before and through the status date represents completed effort; all work beyond it represents remaining effort. Simply put, all dates before the status date are in the past and all dates beyond the status date are in the future. No dates in the past should be estimated; no dates in the future are actual dates.

Unless a status date is provided, the schedule cannot be used to reliably convey past and remaining effort. The status date is an input into the calculations used to update and schedule remaining work. Neither the date on which someone is viewing the schedule nor the latest save date should be used as a substitute for a valid status date (see case study 16).

Best Practice 9: Updating the Schedule Using Actual
Progress and Logic

Updating Duration of Work

Updating duration is the most common method of recording progress because it is the easiest to do. To update activity duration, an actual duration is entered into the plan to record the amount of time (typically working days) elapsed since the last update. If the activity began after the last statusing, an actual start date is entered as well. Next, an updated estimate of time remaining on the activity is entered. The scheduling software calculates percentage complete for the activity based on actual duration and planned remaining duration.[40]

While duration updates are widely used, team members can easily misconstrue them. Because the update applies to the activity duration, it specifically denotes the passage of time from the start date, not actual work accomplished on the activity. For instance, an activity could have used up 80 percent of its scheduled time yet have accomplished only 10 percent of the work. For this reason, it is important that realistic forecasts of remaining duration be updated—while taking into consideration the physical effort remaining to be completed—at the same time as the actual duration is recorded.

Accordingly, recording progress by entering percentage complete is not recommended, because scheduling software adjusts the remaining duration to yield the entered percentage complete. Estimating a percentage of work or time complete is an inexact science, whereas activity managers and schedulers are accustomed to estimating remaining duration. For instance, task managers may confuse inputs and outputs, assuming that if they have worked 7 days on a 10-day activity, they must be 70 percent finished. In addition, asking activity managers for information on actual duration and remaining duration is a

---

[40]Appendix III discusses updating work duration in an EVM environment.

more natural question than requesting a subjective measure of time passed. For example, a response of 80 percent complete on a 5-day activity may indicate that it is almost finished, even if the team has left all the difficult work for the last day.

The subjectivity of percentage complete estimates becomes more apparent the longer the activity's duration. While 25 percent complete may be a viable estimate for a 4-day activity, it is an entirely ambiguous progress measure for a 50-day activity. Finally, percentage complete is based on the expected duration of an activity, which is variable. The exception is in updating level-of-effort activities that represent support activities that are not associated with any discrete product. Level-of-effort activities are typically described as percentage complete by the total duration of the activities they support.

Updating durations generally implies that work performed by a single resource or multiple assigned resources is completed at the same pace and, in terms of percentage complete, is equal to the time spanning the start date and the status date. That is, if the activity is 50 percent complete, then all assigned resources are generally assumed to have completed 50 percent of their work. This is a simplifying assumption that may work well for some program plans.

Other program managers may want to update work by resources to track actual effort by resource unit or group. Tracking actual work progress requires more time and is more complex than updating durations, but it provides more accurate historical data and higher-quality forecasts of remaining effort. Similar to updating durations, updating work progress requires entering actual work performed as well as forecasts for remaining work. The scheduling software then calculates percentage work complete.

Regardless of whether a schedule's progress is marked by duration or work, data should be relevant and should adhere to the definitions set forth in the governing process. Team members and management should have consistent, documented definitions of what constitutes an activity's start, its finish, and its meaningful progress. For example, the actual start date of an activity could be the preparatory research or it could mean only the start of significant work that directly affects the associated product. Likewise, the actual finish date could be the point at which no further work whatsoever is charged to that particular activity or it could simply mean the point at which the activity's successor can begin. In general, opening an activity without the resources to do the work just to record some progress is a poor practice.

Updating the Status of Remaining Work

Time preceding the status date represents history. All unfinished work and activities that have not started should be rescheduled to occur after the status date. If unfinished work remains in the past, the schedule no longer represents a realistic plan for completing the project, and team members will lose confidence in the model. Case study 17 shows an example of activities within a schedule that were not properly updated.

Best Practice 9: Updating the Schedule Using Actual
Progress and Logic

Officials from the Transportation Security Administration's Electronic Baggage Screening Program stated that they had conducted weekly meetings on the schedule and had updated the status accordingly. From the weekly meetings, officials generated weekly reports that identified key areas of concern with regard to schedule shifts and their potential effects on milestones. However, our analysis showed that 30 activities (6 percent of the remaining activities) should have started and finished according to the schedule status date but did not have actual start or actual finish dates.

In addition, the schedule's critical path began in a data collection activity that was not logically linked to any predecessor activities, had a constrained start date, and was marked as 100 percent complete on June 18, 2010—2 months into the future, according to the schedule status date of April 16, 2010.

The schedule also contained 43 activities (9 percent of the remaining activities) with actual start and actual finish dates in the future relative to the schedule status date.

The schedule should be continually monitored to determine when forecasted completion dates differ from the planned dates, which can be used to determine whether schedule variances will affect downstream work.

GAO, *Aviation Security: TSA Has Enhanced Its Explosives Detection Requirements for Checked Baggage, but Additional Screening Actions Are Needed,* GAO-11-740 (Washington, D.C.: July 11, 2011).

PROGRESS RECORDS

Schedules should be statused by reference to progress records for the current time period. Progress records are the sources from which schedulers update the schedule. They are a documentation trail between actual experience on the activity and the progress recorded in the schedule, including actual start and finish dates, the number of resources required, and the amount of work performed to accomplish the activity. Progress records can take many forms according to individual agency or contract requirements. Typically, the document that keeps the activity owner aware of information related to current and upcoming activities is the same form that the owner fills out with actual data and returns to the scheduler for updating (commonly known as a "turnaround" document). This document contains pertinent activity information such as its name, unique ID, original and remaining durations, forecasted and actual start and finish dates, and float. For updating the schedule for the current status period, the progress record should include, at a minimum,

- the actual start date if started or the forecasted start date if not started;
- the actual finish date if finished;
- actual duration or actual work performed; and
- an estimate of the remaining duration or remaining work.

It is important to collect the correct type of data during statusing. Individuals directly managing or performing the work should report progress. If activities are updated by duration, team members should not provide work updates; likewise, if activities are updated by work, team members should not provide duration updates.

Progress records are particularly useful for reviewing remaining duration estimates during attempts to accelerate the schedule (this concept is discussed in Best Practice 10). When the program is completed, progress records provide the historical data necessary for resource, work, and productivity assumptions for future analogous programs. If sufficient attention is paid to recording the way work is actually performed, the resulting archived data will lead to improved accuracy and quality control of similar future programs.

ADDING AND DELETING ACTIVITIES

Activities may be added to represent unplanned work, reflect the actual order of completion of planned work, or refine existing long-duration activities. As with all existing activities, new activities should be given a unique description and should be mapped to the appropriate activity codes. New activities should be reviewed for completeness of predecessor and successor logic, resource assignments, and the effect on the critical path and float calculations. Inserting additional activities may be subjected to the program's schedule change control process.

Activities should not be deleted from a baselined schedule. Deleting an activity may disrupt schedule logic and complicate efforts to compare the current schedule to the baseline. If the activity is no longer valid, its duration should be zeroed out and the activity marked as completed. If some portion of the activity has been completed, the remaining duration of the activity should be zeroed out and a record kept of the completed portion. In addition, a note should be added to the schedule to document why the activity's duration or remaining duration was removed.

OUT-OF-SEQUENCE LOGIC

Rescheduling activities may require adjusting network logic to explain why an activity did not start as planned, particularly if any of its successors have started or been completed. Out-of-sequence logic results from progress on an activity performed in a different order than originally planned such as

- an activity starting before its F–S predecessor has finished;
- an activity starting before its S–S predecessor has started; or
- an activity finishing before its F–F predecessor has finished.

Out-of-sequence progress is common and should be expected, because some activity managers know they can safely start their activities, sometimes challenging their predecessor activity teams to speed up. When out-of-sequence progress occurs, managers and schedulers may choose to retain or override the existing network logic.

Retained Logic

In the case of retained logic, work on the activity that began out of sequence is stopped until its predecessor is completed. As much as possible of the original network logic is preserved because the remainder of the out-of-sequence activity is delayed until the predecessor finishes, to observe its original sequence logic.

Progress Override

In progress override, work on the activity that began out of sequence is permitted to continue, regardless of original predecessor logic. Actual progress in the field supersedes the plan logic, and work on the out-of-sequence activity continues. The predecessor and successor activities may now be executed in parallel, or the original logic on the activities is altered to model their new, more realistic, dependencies. Figure 45 shows the original plan for conducting activities for the interior finishes of the house construction project.

Figure 45: The Original Plan for Interior Finishing

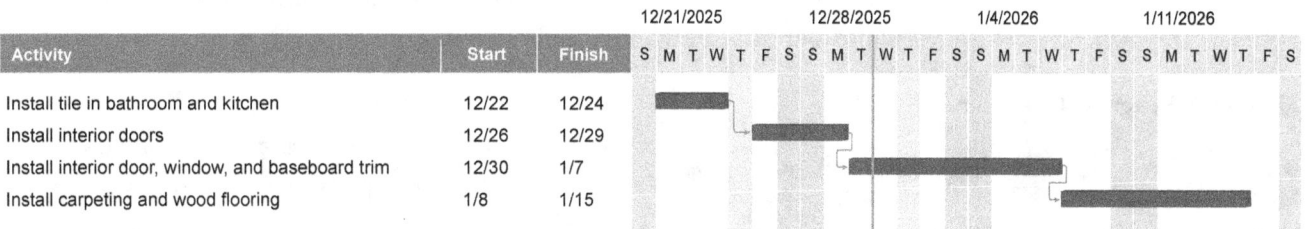

Activity	Start	Finish
Install tile in bathroom and kitchen	12/22	12/24
Install interior doors	12/26	12/29
Install interior door, window, and baseboard trim	12/30	1/7
Install carpeting and wood flooring	1/8	1/15

Source: GAO. | GAO-16-89G

For this example, we assume that the status date is December 30 and that "install tile in bathroom and kitchen" has completed on time. No progress was made on "install interior doors," so its start date is rescheduled for December 31. However, its successor activity "install interior door, window, and baseboard trim" did start on December 29 and two carpenters accomplished 32 hours of work. This is illustrated in figure 46 by the status date (green vertical line) and gray bars that represent actual accomplished effort. If the schedule is statused according to retained logic, the remaining work for "install interior door, window, and baseboard trim" is deferred until the interior doors are installed. In figure 46, 2 days of actual duration are recorded for the "install interior door, window, and baseboard trim" activity, but the remaining duration is rescheduled to begin after "install interior doors" finishes. This start-stop-resume approach may not be an efficient way to install interior trim but it may be important given the reliance of the trimming activity on the installation of interior doors.

Figure 46: Retained Logic

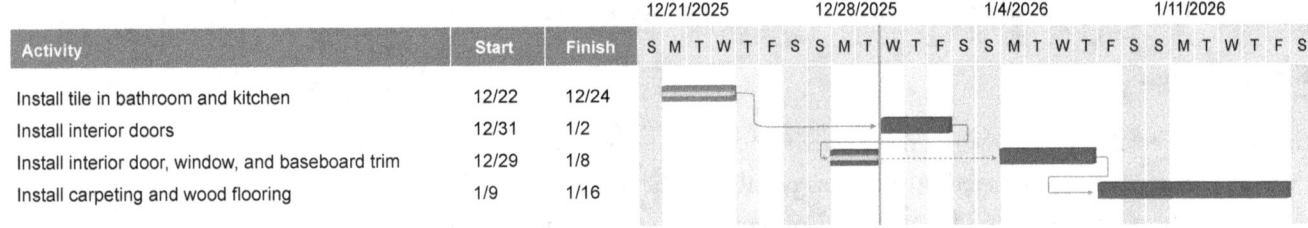

Activity	Start	Finish
Install tile in bathroom and kitchen	12/22	12/24
Install interior doors	12/31	1/2
Install interior door, window, and baseboard trim	12/29	1/8
Install carpeting and wood flooring	1/9	1/16

Source: GAO. | GAO-16-89G

If the schedule is statused according to progress override, the work for "install door, window, and baseboard trim" continues, regardless of the original plan's logic. This concept is illustrated in figure 47. Two days of actual duration are recorded for the trimming activity, and the remaining duration is scheduled to occur in parallel with "install interior doors." While the progress override accelerates the overall construction schedule by 1 working day, the time savings may not be feasible for several reasons. First, it assumes that "install interior door, window, and baseboard trim" does not depend on "install interior doors" finishing as the original plan dictated. Second, it assumes that resources are available to work on both the interior door installation and the interior trim installation. But both activities rely on two carpenters; the concurrent activities now require four carpenters on December 31 and January 2.

Figure 47: Progress Override

Activity	Start	Finish
Install tile in bathroom and kitchen	12/22	12/24
Install interior doors	12/31	1/2
Install interior door, window, and baseboard trim	12/29	1/6
Install carpeting and wood flooring	1/7	1/14

Source: GAO. | GAO-16-89G

Finally, the installation of carpeting and wood flooring, originally scheduled to begin on January 8, is now scheduled to start on January 7. It may be unrealistic to assume that the flooring material and the floor installers will be available 1 day earlier than expected with less than a week's notice. Additionally, the F–S logic between the "install interior doors" and "install door, window, and baseboard trim" activities should be reviewed and repaired. The successor logic of "install interior doors" should be linked to a new correct successor, perhaps "install carpeting and wood flooring." If the logic is not corrected, "install interior doors" will, in effect, have no successor. It will therefore be able to slip to the end of the project without affecting the start date of any successive activity.

Retained logic and progress override are options in scheduling software that must be properly set before updating the status of the schedule. Each approach represents a different philosophy on how to manage unexpected developments in the program, and they can have vastly different effects on forecasted dates. Retained logic is a more conser-

Best Practice 9: Updating the Schedule Using Actual
Progress and Logic

vative approach than progress override and is, in general, the preferred approach. As shown in figure 47, progress override may show better progress, but it may not result in a realistic plan.

It is recommended that the scheduler and activity leads examine each instance of out-of-sequence progress to determine correct responses case by case. While out-of-sequence activities are common, they should nonetheless be reported, and an analysis of the effect on key milestone dates is recommended before out-of-sequence activities are formally addressed. An out-of-sequence activity, despite the selection of "retained logic" or "progress override," degrades the schedule and requires addressing. Regardless of how management wishes to proceed with the work related to the out-of-sequence activity, the logic relationship between activities should be repaired to show the order in which the activities were actually carried out. If left alone, out-of-sequence activities may complicate total float values and cause activities to become artificially critical. More importantly, out-of-sequence activities may represent risk or potential rework, because knowledge or products from the predecessor activity were not complete and available to the successor activity.

VERIFYING STATUS UPDATES AND SCHEDULE INTEGRITY

Once the schedule has been statused, management should review the inputs to verify the updates and assess the effect on the plan. The schedule should be reviewed to ensure the following:

- All activities completed before the status date represent finished work and therefore should have actual start and finish dates. They should be statused as 100 percent complete with actual durations or actual work values.
- In-progress activities should have an actual start date and all work scheduled before the status date is expressed as actual duration or actual effort. All remaining work should be scheduled beyond the status date.
- Activities beyond the status date represent future activities and therefore should not have actual start or actual finish dates. They also should not have actual duration or work values.
- Out-of-sequence activities are addressed purposefully and case by case through either retained logic or progress override.
- Schedule recalculations from changes in estimated work, assigned resources, or duration are verified to ensure meaningful and accurate calculations.
- Resource assignments are assessed for the coming period, and assignments for delayed or out-of-sequence activities are reevaluated for potential overallocations. In addition, resource calendars should be updated to reflect current availability.
- Date constraints are revisited and removed if possible. In particular, soft constraints should be removed if they can no longer affect an activity's start or finish date.

- At least one in-progress activity is critical. If not, it is likely that date constraints or external dependencies are separating successor activities from the in-progress activities.[41] Such breaks in the critical or longest path represent weak or incomplete logic, causing a lack of credibility in the identity of the path and the schedule dates.

Schedule integrity should also be assessed as the schedule is updated. Verifying the integrity of the schedule after each update will ensure that the schedule remains reliable after activities are added, removed, broken down into smaller activities, or sequenced differently from the last period. Common schedule health measures are listed in appendix VI.

It is unlikely in a major program that all activities will be fully identified. As a program changes and completes phases, some activities are overtaken by events and others are generated from lessons learned. Changes made to a schedule, whether minor or major, may need to undergo formal change control according to contract requirements or internal process controls.

The current schedule, once management approves it, should be assigned a version number and archived. This ensures that all status updates can be traced and guarantees that all stakeholders are using the same version of the current schedule.

SCHEDULE NARRATIVE

Regardless of whether a change triggers schedule change control, all changes made to the schedule during statusing should be documented, and salient changes should be justified along with their likely effect on future activities. A schedule narrative should accompany the updated schedule to provide decision makers and auditors a log of changes and their effect, if any, on the schedule time. The scope of the schedule narrative varies with project and contract complexity but should contain, at a minimum,

- the status of key milestone dates, including the program finish date;
- the status of key hand-offs or giver/receiver dates;
- explanations for any changes in key dates;
- changes in network logic, including lags, date constraints, and relationship logic and their effect on the schedule;
- a description of the critical paths, near-critical paths, and longest paths along with a comparison to the previous period's paths; and
- a description of any significant scheduling software options that have changed between update periods, such as the criticality threshold for total float and progress override versus retained logic and whether resource assignments progress with duration.

---

[41]In principle, a critical activity could be scheduled to start the next day after a status update. It would therefore not be in progress at that time, although it would be scheduled to start as soon as possible.

Best Practice 9: Updating the Schedule Using Actual
Progress and Logic

Significant variances between planned and actual performance, as well as actual and planned logic, should be documented and understood. Assessing the updated critical path is particularly crucial. It should be compared to the critical path from the baseline and the prior period's schedule and assessed for reasonableness. Total float should be examined and compared to the last period's schedule to identify trends. For activities that are behind schedule, the remaining duration should be evaluated and the delay's effect on succeeding activities in the network should be understood. If a delay is significant, management should develop a plan to examine the options to recover from the schedule slip. In addition, near-critical paths should be assessed and compared against previous near-critical paths. A decrease in total float values on a near-critical path indicates activities that are slipping. The path may soon be critical, because float will not be available for successive activities on that path. Case study 18 illustrates the effect of a lack of consistent documentation requirements.

Case Study 18: Inconsistent Documentation Requirements, from *Polar-orbiting Environmental Satellites*, GAO-13-676.

GAO conducted a reliability assessment on selected projects of the Joint Polar Satellite System program IMS. We found that the JPSS program office established a preliminary integrated master schedule and implemented multiple scheduling best practices, but the integrated master schedule was not complete and weaknesses in component schedules significantly reduced the program's schedule quality as well as management's ability to monitor, manage, and forecast satellite launch dates. The incomplete integrated master schedule and shortfalls in component schedules stemmed in part from the program's plans to refine the schedule as well as schedule management and reporting requirements that varied among contractors.

The inconsistency in quality of the three schedules had multiple causes. Program and contractor officials explained that in some cases they corrected certain weaknesses with updated schedules. In other cases, the weaknesses lacked documented explanation, in part because the JPSS program office did not require contractors to provide such documentation. We found that schedule management and reporting requirements varied across contractors without documented justification for tailored approaches, which may partially explain the inconsistency of practices in the schedules. Because the reliability of an integrated schedule depends in part on the reliability of its subordinate schedules, schedule quality weaknesses in these schedules transfers to an integrated master schedule derived from them. Consequently, quality weaknesses in JPSS-1 support schedules further constrained the program's ability to monitor progress, manage key dependencies, and forecast completion dates.

GAO, *Polar Weather Satellites: NOAA Identified Ways to Mitigate Data Gaps, but Contingency Plans and Schedules Require Further Attention*, GAO-13-676 (Washington, D.C.: September 19, 2013).

Reporting and Communication

As noted earlier, a schedule is a fundamental program management tool that specifies when work will be performed in the future and how well the program is performing against an approved plan. It is therefore particularly important that all stakeholders be able to view information stored in the schedule related to their specific roles and needs to successfully manage and execute the plan. Stakeholders include, among others, program team members, activity managers, government customers, resource managers, subcontractors, program sponsors, finance specialists, and decision makers. Each stakeholder requires a different level and type of information that depends on whether the stakeholder is internal to the program or external. Reports can encompass actual data (status), actual versus planned data (progress), and predictive data (forecasts). A well-constructed, comprehensive schedule is a database that contains actual, planned, and forecast activity as well as resource and cost information. It can report reliable data quickly at all levels of detail.

The regularity of schedule reporting and the level of detail that is reported naturally varies by stakeholder and project complexity. At the level of senior decision maker, high-level summary trend charts and key milestone schedules reporting monthly or quarterly progress are most useful. These reports typically include progress and forecast information on contractual and deliverable milestones and major program phases, as well as summary critical path and contingency information. High-level trend information is also useful, such as key milestone completions, contingency burn rate, and resource availability. In addition, the level of detail depends on the complexity and risk of certain WBS items. For example, a series of complex activities on the critical path may be reported to program management in more detail than less complex, noncritical activities.

Best Practices Checklist: Updating the Schedule Using Actual Progress and Logic

- Schedule progress is recorded regularly and the schedule has been updated recently. Schedule status is updated with actual and remaining progress.

- Schedule status is based on progress records for the current time period; they include pertinent activity information such as name, unique ID, original and remaining durations, forecasted and actual start and finish dates, and float.

- The status date (or data date) denoting the date of the latest update to the schedule is recorded.

- At least one in-progress activity is critical.

- No activities precede the status date without actual start or finish dates and actual effort up to the status date. No activities beyond the status date have actual start or finish dates or actual effort.

Best Practice 9: Updating the Schedule Using Actual
Progress and Logic

- Activities that are behind schedule by the status date have a remaining duration estimate, and the delay's effect has been assessed.

  - If the delay is significant, plans to recover the implied schedule slip have been evaluated and implemented, if so decided.
  - Resources are reviewed and may be reassigned, depending on schedule progress.

- Responsibility for changing or statusing the schedule is assigned to someone who has the proper training and experience in CPM scheduling.

- Changes that were made to the schedule during the update have been documented.

- New activities are reviewed for completeness of predecessor and successor logic, resource assignments, and effects on the critical path and float calculations.

- Activities that have started out of sequence or have been completed out of sequence have been addressed using either retained logic or progress override to reflect the order in which they were carried out.

- Management reviews schedule updates to verify and assess effects on the plan. Significant variances between planned and actual performance, as well as between planned and actual logic, are documented and understood.

- The schedule structure is examined after each update to ensure that logic is not missing or broken, all date constraints are necessary, and no artifacts impede the ability of the schedule to dynamically forecast dates.

- The current schedule, once management approves it, is assigned a version number and archived.

- A schedule narrative accompanies each status update and includes

  - the status of key milestone dates, including the program finish date;
  - the status of key hand-offs or giver and receiver dates;
  - explanations for any changes in key dates;
  - changes in network logic, including lags, date constraints, and relationship logic and their effect on the schedule;
  - a description of the critical paths, near-critical paths, and longest paths along with a comparison to the previous period's paths; and
  - any significant scheduling software options that have changed between update periods, such as the criticality threshold for total float; progress override versus retained logic; or whether resource assignments progress with duration.

BEST PRACTICE 10

Maintaining a Baseline Schedule

Best Practice 10: A baseline schedule is the basis for managing the program scope, the time period for accomplishing it, and the required resources. The baseline schedule is designated the target schedule and is subjected to a configuration management control process. Program performance is measured, monitored, and reported against the baseline schedule. The schedule should be continually monitored so as to reveal when forecasted completion dates differ from baseline dates and whether schedule variances affect downstream work. A corresponding basis document explains the overall approach to the project, defines custom fields in the schedule file, details ground rules and assumptions used in developing the schedule, and justifies constraints, lags, long activity durations, and any other unique features of the schedule.

BASELINE AND CURRENT SCHEDULES

Establishing a baseline schedule is essential to effective management. A baseline schedule represents the original configuration of the program plan and signifies the consensus of all stakeholders regarding the required sequence of events, resource assignments, and acceptable dates for key deliverables. It is consistent with both the program plan and the program budget plan and defines clearly the responsibilities of program performers. The baseline schedule includes not only original forecasts for activity start and finish dates but also the original estimates for work, resource assignments, critical paths, and total float.

The baseline schedule is not the same as the current schedule. The current schedule is updated from actual performance data, as described in Best Practice 9. Therefore, it is the latest depiction of performance and accomplishments, along with the latest forecast of remaining dates and network logic. The baseline schedule represents the program's commitments to all stakeholders, while the current schedule represents the actual plan to date.

The current schedule is compared to the baseline schedule to track variances from the plan. Deviations from the baseline inform management that the current plan is not following the original plan all stakeholders have agreed to. Deviations imply that the current approach to executing the program needs to be altered to align the program to the original plan or that the plan from this point forward should be altered.

Comparing the current status of the schedule to the baseline schedule can help managers identify the cause of the deviation, thereby allowing them to target specific areas such as resource assignments, network logic, and other factors for immediate mitigation. Without a formally established baseline schedule to measure performance against, management lacks the ability to identify and mitigate the effects of unfavorable performance.

The final version of the current schedule—the "as-built" schedule—represents the plan as executed to completion. Particular care should be taken to archive this final version. Once the project has been completed, the as-built schedule becomes a database of the actual sequence of events, activity durations, required resources, and resource productivity. These can be compared to the original plan for an assessment of lessons learned, and the data become a valuable basis of estimate input for schedule estimates of analogous projects.

As-built schedules are also useful for creating and validating fragmentary networks, or "fragnets." A fragnet is a subordinate network that represents a sequence of activities typically related to repetitive effort. Subordinate networks can be inserted into larger networks as a related group of activities. For example, a related group of activities may occur for each systems test, regardless of the actual product. In this case, a fragnet related to systems test, representing a well-known sequence of events and expected duration, may be inserted into various product schedules.

The baseline should be set promptly after a program begins. A schedule baseline is typically in place between 3 and 6 months of contract award, although the timing depends on contract size and type, requirements, and risk.[42] Projects should operate on an interim schedule until the schedule baseline is in place. The level of detail in the baseline schedule also depends on contract type, industry type, risk, and agency guidelines. For example, management may choose to baseline the entire detailed IMS or only intermediate-level summary activities leading to key milestones. The greater the baselined detail is, the greater will be the understanding of performance, variances, forecasts, and assessed effects of potential changes—yet this must be balanced with the time necessary to formally approve, change, and track at that level.

In addition, baseline creation and approval may take place in concert with the program's rolling wave process. That is, as periods of summary and intermediate-level planning packages are planned in greater detail, the baseline is updated to reflect that detail. The intent is that once formally approved and archived, the baseline schedule reflects the agency's commitment to allocating resources and becomes the basis against which actual performance and accomplishments can be measured, monitored, and reported. The parties should agree that the baseline refers to the maintenance of a schedule for meeting contractual deliverable and program control milestones. It typically does not constitute a strict adherence to estimates of activity durations, resource assignments, or logic.

_____

[42] Establishing and maintaining a schedule baseline is highly significant when EVM is a requirement. Appendix III has more information.

BASIS DOCUMENT

The accuracy of the IMS as a model for the program depends on the mutual understanding of all stakeholders of the schedule's structure, use, and underlying assumptions. Thorough documentation is essential for validating and defending a baseline schedule. A well-documented schedule can present a convincing argument for a schedule's validity and can help answer decision makers' and oversight groups' probing questions. A well-documented schedule is essential if an effective independent review is to ensure that it is reliable.

A schedule basis document is a single document that defines the organization of the IMS, describes the logic of the network, describes the basic approach to managing resources, and provides a basis for all parameters used to calculate dates. As noted in Best Practice 1, the government program management office is responsible for integrating all government and contractor work into one comprehensive program plan. Therefore, the government program management office creates and approves the schedule basis document once the schedule baseline is approved. While the basis document is not updated as often as the schedule narrative described in Best Practice 9, it should be considered to be a living document in that it reflects updates to the baseline schedule. Both the baseline schedule and basis document should be referenced and updated in accordance with the established change process as the program progresses. At a minimum, the basis document should include the following.

- A program overview or a general description of the program, including general scope and key estimated life-cycle dates, and descriptions of each stakeholder, including the program owner, prime contractor and subcontractors, and key stakeholder agencies.
- A general description of the overall execution strategy, including the type of work to be performed, and contracting and procurement strategies.
- A description of the overall structure of the IMS, including the scope and purpose of projects, staff responsible for each project, the relationship between projects, a WBS dictionary, the status delivery dates for each project, and a list of key hand-off products and their estimated dates. In addition, it refers to relevant contract data requirements lists and data item descriptions associated with schedules.[43]
- A description of the settings for key options for the relevant software, such as criticality threshold, progress override contrasted with retained logic, and the calculation of critical paths and whether work progresses with duration updates.
- A definition and justification in the basis document of all ground rules and assumptions used to develop the baseline, including items specifically excluded from the schedule. If rolling wave planning is used, key dates or milestones for detail and planning periods are defined.

---

[43]A contract data requirements list is part of a procurement contract. A data item description describes the standardized format and instructions for preparing and delivering the data.

- A justification of each use of a lag and date constraint that is clearly associated with the activities that are affected. Missing successor or predecessor logic is also justified.
- The documentation of each project, activity, and resource calendar, along with a rationale for workday exceptions (holidays, plant shutdowns, and the like), working times, and planned shifts.
- An appropriately detailed explanation or rationale for the basic approach to estimating key activity durations and justification of the estimating relationship between duration, effort, and assigned resource units.
- A justification of each use of a long-duration activity in the near-term.
- A description of how LOE activities are identified and statused.
- Definitions of all acronyms and abbreviations.
- A table of the purposes of custom fields.
- A description of resources used in the plan and the basic approach for updating resource assignments, along with schedule updates and average and peak resource demand projections. The document defines all resource groups used in the schedule and justifies planned overallocations in appropriate detail. In addition, key material and equipment resources are described in the context of their related activities.
- A description of critical paths, longest paths, and total float. A discussion of the critical paths, including justification for the criticality threshold, rationale for single or multiple critical paths, and justification for any gaps in a noncontinuous critical path. If the schedule has date constraints, a description of the longest path is included. In addition, abnormally large amounts of total float are justified.
- A description of the critical risks as prioritized in a quantitative schedule risk analysis, along with a discussion of the appropriate schedule contingency.
- A detailed description of the updating and schedule change management processes. Any additional documentation that governs them is referenced.

Thorough documentation helps with analyzing changes in the program schedule and identifying the reasons for variances between estimates and actual results. In this respect, basis documentation contributes to the collection of cost, schedule, and technical data that can be used to support future estimates.

THE CHANGE PROCESS

The baseline and current schedules should be subjected to a change control process. A control process governs when and how technical and programmatic changes are applied to the baseline, as well as the content of the current IMS. A proper change control process helps ensure that the baseline and current schedule are accurate and reliable. Without a documented, consistently applied schedule change control process, program staff might continually revise the schedule to match performance, hindering management's insight into the true performance of the project. Undocumented or unapproved changes hamper performance measurement and may result in inaccurate variance reporting, inconsistent stakeholder versions of the plan, and unreliable schedule data. Moreover, if changes are not controlled and fully documented, performance cannot be accurately measured against the original plan.

Program management decides on the extent to which the IMS is subjected to a change control process; it depends on the program size, scope, risk, and complexity. Less complex projects may subject only key milestones or high-level WBS activities to the change control process, so that changes to supporting detailed activities do not require change approval. For more complex and tightly controlled schedules, lower-level detail activities as well as network dependencies may also be subjected to change approval.

The current schedule, once management approves it, should be assigned a version number and archived. This ensures that all status updates can be traced and it guarantees that all stakeholders are using the same version of the current schedule. Case study 19 highlights a complex, highly detailed schedule that was successfully updated regularly according to a defined process.

Case Study 19: Baseline Schedule, from *2010 Census*, GAO-10-59

For the 2010 Decennial Census schedule, the U.S. Census Bureau had implemented processes around its master schedule that complied with a number of scheduling process criteria that are important to maintaining a schedule that is a useful management tool. The Bureau ensured that the schedule would be complete with input from stakeholders throughout the agency, with reviews of previous schedules that built on a number of census tests during the decade. As a result, the schedule was the primary source senior managers used to determine weekly what census activity was ahead of or behind schedule, and it provided a resource for determining the effects on the overall project of any delays in major activities.

The Bureau had documented and implemented a formal process for keeping the data in the schedule current. Staff within each Bureau division were responsible for ensuring that the status of schedule activities was updated weekly. Staff updated the actual start and finish dates, the percentage of each activity completed so far, and estimates of the time remaining to complete each activity in progress. The Bureau was recording status information on an average of more than 1,300 activities in the schedule during any given week, generating historical data that could provide valuable input to future schedule estimates.

In addition, the Bureau implemented a formal change control process that preserved a baseline of the schedule so that progress could be meaningfully measured. The Bureau's criteria for justifying changes were clearly documented, approval for them was required by a team of senior managers, and each team acknowledged their effect within the Bureau. About 300 changes had been approved since the master schedule was baselined in May 2008. Even corrections to the schedule for known errors, such as incorrect links between activities, had to be approved through the change control process, helping to ensure the integrity of the schedule.

GAO, *2010 Census: Census Bureau Has Made Progress on Schedule and Operational Control Tools, but Needs to Prioritize Remaining System Requirements,* GAO-10-59 (Washington, D.C.: November 13, 2009).

Changes to the baseline schedule should be limited to revisions for new or reduced scope or for formal replanning. At times, however, management may conclude that the remaining schedule target for completing a program is significantly insufficient and that the current baseline is no longer valid for realistic performance measurement. The purpose of schedule rebaselining is to restore management's control of the remaining effort by providing a meaningful basis for performance management.

Working to an unrealistic baseline could worsen an unfavorable schedule's condition. For example, if variances become too big, they may obscure management's ability to discover newer problems that could still be resolved. To quickly identify new variances, a schedule rebaseline normally eliminates historical variances. A rebaselined schedule should be rare. If a program is rebaselined often, it may be that the scope is not well understood or simply that program management lacks effective discipline and is unable to develop realistic estimates.

BASELINE ANALYSIS

Schedules deviate from the baseline as a program is executed. Changes in resource availability, late or early key deliveries, unexpected additional work activities, and risks can contribute to deviation. Although they are often perceived as something bad, negative variances provide valuable insight into program risk and its causes. Positive variances can indicate problems as well. For example, early starts may cause issues with out-of-sequence logic and can disrupt the scheduling of future resources.

Understanding the types of activities that have started earlier or later than planned is vital as well. For instance, positive variances may not be desirable if only relatively easily accomplished activities are completed early while critical activities are delayed. Variances empower management to decide how best to handle risks. Schedule deviations from the baseline plan give management at all levels information about whether corrections will bring the program back on track or completion dates need updating.

A schedule variance does not necessarily mean program delay; it means that work was not completed as planned. Negative schedule variances should be investigated to see if the effort is on the critical path. If it is, then the whole program will be delayed. In addition, activities that vary significantly from their baseline may create a new critical path or near-critical path.

Carefully monitoring the schedule allows for quickly determining when forecasted completion dates differ from the baseline dates. In this respect, progress can be evaluated for whether it has met planned targets. Activities may be resequenced or resources realigned. It is also important to determine whether schedule variances are affecting successive work activities. For example, a schedule variance may compress remaining activities' durations or cause "stacking" of activities toward the end of the program, to the point at which it is no longer realistic to predict success.

A schedule Gantt chart can be used to show baseline dates, forecast dates, and actual progress for each activity. Consider the activities in the foundation and underground work phase of the house construction project in figure 48. Red activities are forecasted to be critical and blue activities will have 4 days of total float on their paths. The gray bars below the red and blue activity bars represent the baseline start and finish dates. The original plan is to begin staking on September 15 and to complete pouring footings and pads by September 24. Before any progress is made on the project, baseline dates and forecast dates are the same.

Figure 48: Baselined Activities

Activity	Start	Actual start	Finish	Total float	9/14/2025 / 9/21/2025
Lay out and stake property and excavation	9/15	NA	9/15	0 days	
Dig foundation and basement	9/16	NA	9/16	0 days	
Lay out and form footings and pier pads	9/17	NA	9/19	0 days	
Excavate for and install underground sewer	9/16	NA	9/16	4 days	
Inspect underground sewer	9/17	NA	9/17	4 days	
Install footing and pier rebar	9/22	NA	9/22	0 days	
Inspect footing and pier rebar	9/23	NA	9/23	0 days	
Pour footings and pads	9/24	NA	9/24	0 days	

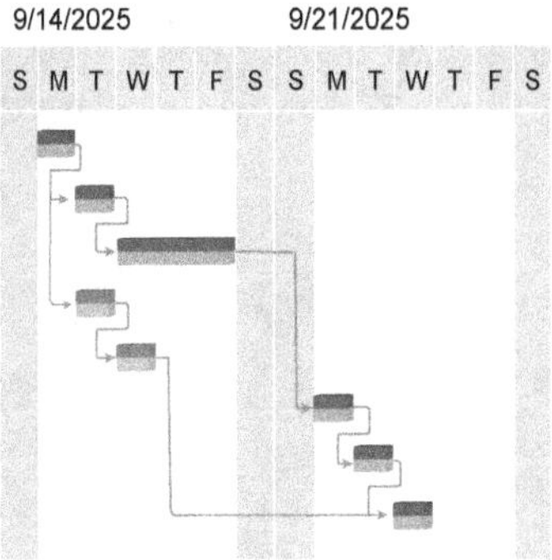

Source: GAO. | GAO-16-89G

Figure 49 gives the current schedule for the same plan, now statused through September 19. The green vertical line represents the current status date, and the gray horizontal bars represent progress on each activity.

Figure 49: Updated Status Compared with Baseline

Activity	Start	Actual start	Finish	Total float	9/14/2025 / 9/21/2025
Lay out and stake property and excavation	9/15	9/15	9/15	0 days	
Dig foundation and basement	9/16	9/16	9/16	0 days	
Lay out and form footings and pier pads	9/17	9/17	9/19	0 days	
Excavate for and install underground sewer	9/19	9/19	9/23	0 days	
Inspect underground sewer	9/24	NA	9/24	0 days	
Install footing and pier rebar	9/22	NA	9/22	1 day	
Inspect footing and pier rebar	9/23	NA	9/23	1 day	
Pour footings and pads	9/25	NA	9/25	0 days	

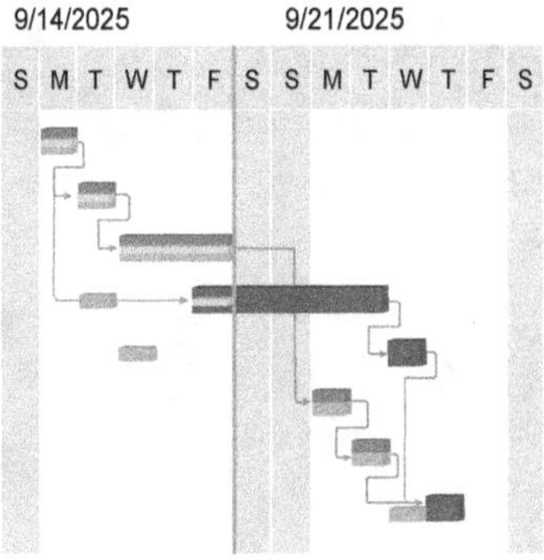

Source: GAO. | GAO-16-89G

As seen in the current schedule, the "layout and stake property and excavation," "dig foundation and basement," and "lay out and form footings and pier pads" activities started and finished on time. However, not only did the "excavate for and install underground sewer" activity start 3 days late but also its effort is expected to take 32 hours longer than originally planned. The 3-day late start and the extra effort consume the available 4 days of total float on the path, and the activities "excavate for and install underground sewer" and "inspect underground sewer" are now critical. The activities "install footing and pier rebar" and "inspect footing and pier rebar" activities are no longer critical, because the delay in the underground sewer work created a day of total float for those activities. The start variance of "excavate for and install underground sewer" is 3 days and the finish variance is expected to be 5 days. The effect of the underground sewer delay can clearly be seen in the Gantt chart. Visually, the activity bars are far removed from their original baselined positions in the timeline.

In this situation, the general contractor has several options to get the construction schedule back on track:

- Adjust the working calendar. Installation could take place over the weekend. Alternatively, the plumbers could work overtime to finish installing the sewer by the morning of September 23. Both options assume that resources are available for overtime work on short notice and that the project can afford the additional labor rate associated with overtime. It also assumes there will be no breach of local law such as restrictions on construction noise.
- Increase resources. Adding additional workers to the sewer installation activity would reduce the activity's duration by 1 day. Again, this assumes that additional resources are available on short notice. The additional labor cost is unavoidable because of the additional work required, but this option leaves the labor cost at the normal rate.
- Perform activities concurrently. Overlapping work is typically an option but not in this case. The underground sewer work must be finished before it can be inspected.
- Do nothing and allow pouring of the footing and pads a day late. This assumes that the negative float of 1 day will be addressed at some point beyond the foundation and underground work phase, perhaps by adding resources during framing or finishing. However, this option affects many subsequent resource assignments such as carpenters, plumbers, electricians, HVAC specialists, and bricklayers. Each assignment will have to be rescheduled if the delay is carried beyond foundation and underground work.

Figure 50 shows the effect of the general contractor's plan to get the program back on track by adding extra resources to the installion activity. This reduces the installation activity by 1 day and places the "pour footings and pads" activity back on its originally planned date.

Figure 50: Updated Status and Proposed Plan Compared with Baseline

Activity	Start	Actual start	Finish	Total float	
Lay out and stake property and excavation	9/15	9/15	9/15	0 days	
Dig foundation and basement	9/16	9/16	9/16	0 days	
Lay out and form footings and pier pads	9/17	9/17	9/19	0 days	
Excavate for and install underground sewer	9/19	9/19	9/22	0 days	
Inspect underground sewer	9/23	NA	9/23	0 days	
Install footing and pier rebar	9/22	NA	9/22	0 days	
Inspect footing and pier rebar	9/23	NA	9/23	0 days	
Pour footings and pads	9/24	NA	9/24	0 days	

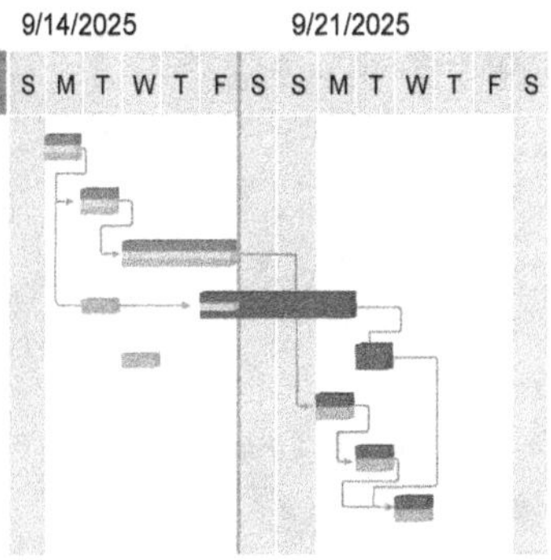

Source: GAO. | GAO-16-89G

The threshold for reporting variances varies by program size, complexity, and risk. All stakeholders should agree on the threshold and it should be formally defined in a governing document. In particular, guidance should take into account the threshold for the number of days the activity is delayed as well as the available float. If a variance exceeds the threshold, it should be reported to management along with a detailed description that includes the cause and recommended corrective actions. Note that because level-of-effort activities support work activities, they never show a variance.

Various schedule measures should be analyzed to identify and better understand the effect of schedule variances. Some examples of measures for comparing the current schedule to the baseline include the number of activities that

- have started early, on time, and late;
- have finished early, on time, and late;
- should have started but have not;
- should have finished but have not;
- should not have started but have;
- should not have finished but have.

It is important to note that the validity of variances is directly related to the reliability of the schedule. If the schedule is not well constructed, comprehensive, or credible, any variances resulting from comparing actual status to the baseline schedule will be questionable.

TREND ANALYSIS

Trend analysis involves collecting schedule measurement data during each status period and plotting them over time. Sustained variances in one direction should be documented to assess whether the program is heading toward a problem and to determine whether action is necessary. Trend analysis provides valuable information about how a program is performing. Knowing what has caused problems in the past can help determine whether they will continue. Typical schedule measure data that can help managers know what is happening in their programs are

- comparisons of the cumulative number of completed activities to the cumulative number of baselined activities planned to be completed;
- comparisons of the cumulative number of activities forecasted to be completed to the number of baselined activities planned to be completed;
- available resources over time, highlighting overallocations;
- total float and schedule contingency tracked by status period;
- comparisons of actual, forecasted, and baselined completed activities.

The use of total float over time is a key measure for monitoring the health of the program. A quick draw down of total float early in the program may indicate that more serious schedule delays and cost overruns will occur in the future.

Comparisons of actual, forecasted, and baselined completed activities are useful for identifying unrealistic peaks of activity. Peaks of activity will have to be completed at a higher rate than the rate at which activities have previously completed. For example, if on average 15 detail activities have been accomplished per 2-week period yet forecasts imply that 30 detail activities need to be completed every 2 weeks to finish on time, the schedule is most likely unrealistic.

A measure known as the baseline execution index (BEI) is commonly used in baseline analyses. It is the ratio of the number of detail activities that were completed to the number of detail activities that should have been completed by the status date. A BEI of 1 indicates that the project is performing according to plan. A BEI less than 1 indicates that, in general, fewer activities are being completed than planned; a BEI greater than 1 indicates that, in general, more activities are being completed than planned.

The BEI is an objective measure of overall schedule efficiency because it compares actual completions to planned completions. However, it is a summary measure—that is, it does not provide insight into why activities are not being completed according to plan or take into account the importance of their not being completed according to plan. For example, delayed activities that are on the critical path or on near-critical paths are given weight equal to delayed activities that have free float available.

To provide additional insight, the BEI can be calculated against any group of activities—WBS level, resource group, or activity duration. It can also be calculated for different periods of time—for example, from the start of the program through the status date or for the most recent reporting period. The BEI is always 1 at the end of a project if, eventually, all activities are completed. Finally, if contingency activities are included in the schedule network, their durations should be zeroed out before calculating the BEI. Including contingency activities can artificially lower the BEI because they have no actual start or finish dates.

STRATEGIES FOR RECOVERY AND ACCELERATION

Table 6 summarizes common techniques that schedulers and management teams use to recover schedule variances and, in general, accelerate a schedule. Schedule recovery is a managerial effort to recover from the interruption of events on the planned schedule. Acceleration refers to a reduction in the duration of the overall schedule. The focus for schedule recovery or acceleration should be on critical activities and, in particular, on long-duration critical activities. Because critical activities drive key milestone dates, the overall program duration will be reduced only by reducing the critical paths. Some of the strategies described in table 6 may have contractual implications.

Table 6: Strategies for Recovery and Acceleration

Technique	Description	Side effects
Crashing	Add resources to time-dependent activities to complete work faster	Requires additional resources and thus increases costs; may also reduce quality if activities are executed faster and with less-experienced labor
Fast tracking	Reduce the sequential dependencies between activities to partial dependencies. For example, F–S logic is reduced to S–S logic to force parallel work	Resources may become overallocated; quality may also be reduced and risk introduced if activities ideally executed in sequence are now executed in parallel
Split long activities	Split long activities into shorter activities that can be worked in parallel	Resources may become overallocated
Review constraint and lag assumptions	Reassess assumptions related to forcing activities to begin on certain dates	If the original date constraint or lag is justified, removing the constraint or lag may not be realistic
Review duration estimates	Revisit duration estimates using progress records as actual effort is recorded	
Add overtime and reduce vacations	Review nonworking periods and assign overtime work	Costs will increase over standard labor rates; as overtime increases, morale decreases, eventually affecting the quality of the product negatively
Reduce scope	Decrease scope to reduce both duration and costs	Scope is the primary reason for performing the work, and it may not be possible to delete some requirements
Schedule contingencies	Allocate contingency to absorb delay in accordance with identified risk mitigation plans	Using contingency too early in a project reduces the likelihood of the program's completing on time, particularly if the reason for a delay was a risk that was not previously identified and quantified

Source: GAO and NDIA | GAO-16-89G.

Regardless of what combination of methods is used for acceleration or recovery of variances, the schedule should be archived before the change. After the schedule is modified, the techniques used should be documented and the critical path recalculated and assessed for reasonableness.

Best Practices Checklist: Maintaining a Baseline Schedule

- A baseline schedule

  - is set promptly after the program begins;
  - is the basis for measuring performance;
  - represents the original configuration of the program plan and signifies the consensus of all stakeholders regarding the required sequence of events, resource assignments, and acceptable dates for key deliverables;
  - is compared to the current schedule to track variances from the plan.

- The as-built version of the schedule is planned to be archived and will

  - represent the plan as executed to completion;
  - be compared to the original plan for an assessment of lessons learned;
  - become a valuable basis of estimate input for schedule estimates of future analogous projects;
  - become the basis for creating and validating fragnets where possible.

- A schedule basis document is a single document that

  - defines the organization of the IMS;
  - describes the logic of the network;
  - describes the basic approach to managing resources;
  - provides a basis for all parameters used to calculate dates;
  - describes the general approach to the project;
  - defines how to use the schedule file;
  - describes the schedule's unique features;
  - describes the schedule change management process;
  - contains a dictionary of acronyms and custom fields;
  - gives an overview of the assumptions and ground rules, including justification for
    - calendars,
    - lags,
    - date constraints,
    - long activity durations, and
    - calendars and working times;
  - describes the use and assignments of resources within the schedule in appropriate detail;
  - describes the critical risks prioritized in a schedule risk analysis as well as schedule contingency;
  - discusses the derivation of the critical paths, longest path, and total float for the program;
  - is considered a living document that reflects updates to the baseline schedule.

Best Practice 10: Maintaining a Baseline Schedule

- Changes to the baseline schedule are reviewed and approved according to the change control process.

- Schedule variances that exceed predetermined thresholds are reported to management, along with the cause and any corrective actions.

- Negative schedule variances are investigated to see if the associated effort is on the critical path.

- Schedule measures, such as the number of activities that have started or finished early, on time, or later than planned, are analyzed for the effect of any variances.

- Trend analysis is conducted regularly to examine measures such as decreasing float and schedule contingency erosion.

- The focus of any schedule recovery or acceleration techniques is on critical activities.

FOUR CHARACTERISTICS OF A RELIABLE SCHEDULE

GAO's research tells us that the four characteristics of a high-quality, reliable schedule are that it is comprehensive, well-constructed, credible, and controlled. A comprehensive schedule includes all activities for both the government and its contractors necessary to accomplish a program's objectives as defined in the program's WBS. The schedule includes the labor, materials, travel, facilities, equipment, and the like needed to do the work and depicts when those resources are needed and when they will be available. It realistically reflects how long each activity will take and allows for discrete progress measurement.

A schedule is well-constructed if all its activities are logically sequenced with the most straightforward logic possible. Unusual or complicated logic techniques are used judiciously and justified in the schedule documentation. The schedule's critical path represents a true model of the activities that drive the program's earliest completion date, and total float accurately depicts schedule flexibility.

A schedule is credible if it is horizontally traceable—that is, it reflects the order of events necessary to achieve aggregated products or outcomes. It is also vertically traceable: activities in varying levels of the schedule map to one another and key dates presented to management in periodic briefings are in sync with the schedule. Data about risks are used to predict a level of confidence in meeting the program's completion date. Necessary schedule contingency and high-priority risks are identified by conducting a robust schedule risk analysis.

Finally, a schedule is controlled if trained schedulers update it regularly using actual progress and logic—based on information provided by activity owners—to realistically forecast dates for program activities. Updates to the schedule are accompanied by a schedule narrative that describes salient changes to the network. The current schedule is compared against a designated baseline schedule to measure, monitor, and report the program's progress. The baseline schedule is accompanied by a basis document that explains the overall approach to the program, defines ground rules and assumptions, and describes the unique features of the schedule. The baseline schedule and current schedule are subjected to configuration management control. Table 7 shows how the 10 scheduling best practices can be mapped to these four characteristics.

Table 7: Best Practices Entailed in the Four Characteristics of a Reliable Schedule

Schedule characteristic	Best practice
Comprehensive, reflecting • all activities as defined in the program's WBS • labor, materials, travel, facilities, equipment, and the like needed to do the work and whether those resources will be available when needed • how long each activity will take, allowing for discrete progress measurement with specific start and finish dates	1. Capturing all activities 3. Assigning resources to all activities 4. Establishing the durations of all activities
Well constructed, with • all activities logically sequenced with predecessor and successor logic • limited and justified use of unusual or complicated logic • a critical path that determines the activities that drive the program's earliest completion date • total float that accurately reflects the schedule's flexibility	2. Sequencing all activities 6. Confirming that the critical path is valid 7. Ensuring reasonable total float
Credible, reflecting • the order of events necessary to achieve aggregated products or outcomes • varying levels of activity, supporting activity, and subtasks • a level of confidence in meeting a program's completion date based on data about risks for the program • necessary schedule contingency and prioritized risks based on a robust schedule risk analysis	5. Verifying that the schedule can be traced horizontally and vertically 8. Conducting a schedule risk analysis
Controlled, being • updated regularly by schedulers trained in critical path method scheduling • statused using actual progress and logic to realistically forecast dates for program activities • accompanied by a schedule narrative that describes updates to the current schedule • compared against a baseline schedule to determine variances from the plan • accompanied by a corresponding basis document that explains the overall approach to the program, defines assumptions, and describes unique features of the schedule • subject to a configuration management control process	9. Updating the schedule using actual progress and logic 10. Maintaining a baseline schedule

Source: GAO | GAO-16-89G.

APPENDIX I

OBJECTIVES, SCOPE, AND METHODOLOGY

Our approach to developing this guide was to ascertain best practices from leading practitioners and to develop standard criteria to determine the extent agency programs and projects meet industry scheduling standards. Each best practice was developed and validated in consultation with a committee of cost estimating, scheduling, and earned value analysis specialists. These specialists meet at GAO headquarters semi-annually. The meetings are open to all with interest and expertise in cost estimating, schedule, and earned value management, as well as program managers and agency executives. Meeting members are from government agencies, private companies, independent consultant groups, trade industry groups, and academia from around the world. Agendas are sent to a mailing list of approximately 600 experts, and we receive feedback and discussion on agenda items through the meeting discussion and from telephone participants and email from members. Meeting minutes are extensively documented and archived.

Consistent with our methodology in the formulation of the GAO *Cost Estimating and Assessment Guide* (GAO-09-3SP), we released a public exposure draft of the GAO *Schedule Assessment Guide* in May 2012 and sought input and feedback from all who were interested for 2 years. We vetted each comment we received on whether it was actionable, within scope, technically correct, and feasible. We received nearly 300 comments on the guide before we released it for public exposure and over 1,000 comments during the public exposure period. We received comments from the public, private companies, trade industry groups, and university researchers and extensive comments from government agencies and government working groups.

We compared the standards detailed in the guide with schedule standards and best practices other agencies and organizations had developed. We found that our standards are comparable to them, with limited exceptions. A comparison of our standards to other sets of standards is in appendix VII.

We conducted our work from November 2010 to November 2015 in accordance with all sections of GAO's Quality Assurance Framework that are relevant to our objectives. The framework requires that we plan and perform the engagement to obtain sufficient and appropriate evidence to meet our stated objectives and to discuss any limitations in our work. We believe that the information and data obtained, and the analysis conducted, provide a reasonable basis for the guidance in this product.

APPENDIX II

An Auditor's Key Questions and Documents

Best Practice 1: Capturing All Activities

Key Questions

1. Is there an IMS for managing the entire program (not just a block, increment, or prime contractor)? Is the schedule defined at an appropriate level to ensure effective management?

2. Is the IMS maintained in scheduling software and linked to external, detailed project schedules?

3. How does management ensure the accuracy of reported schedule information? Do the government program management office and contractors have different scheduling software systems? If so, how is integrity preserved and verified when converting the schedule?

4. Does the IMS include government, contractor, and applicable subcontractor effort?

5. Does the schedule reflect the program WBS and does the WBS allow tracking key deliverables? Does every activity trace to an appropriate WBS element, and do the activities define how the deliverables will be produced? Does the schedule WBS map to the cost estimate WBS? Is there a WBS dictionary?

6. Are key milestones identified and are they consistent with the contract dates and other key dates management established in the baseline schedule?

7. Does the schedule have clear start and finish milestones? Are there too many milestones in relation to detail activities?

8. Are activities within the schedule easily traced to key documents and other information through activity or task codes? Are all contractor activities mapped to the contract statement of work (SOW) to ensure that all effort is accounted for in the schedule?

9. Are activity names unique and descriptive? Are activities phrased in verb-noun combinations (for example, "develop documentation")? Are milestones named with verb-noun or noun-verb combinations (for example, "start project" or "project finished")?

10. Are level-of-effort activities clearly marked?

11. Does the schedule include significant risk mitigation efforts as discrete activities? If not, how are they documented and tracked?

Key Documentation

1. Work breakdown structure (WBS) and dictionary

2. Statement of work (SOW), integrated master plan (IMP) and mission requirements, as applicable

3. SOW crosswalk to the WBS and schedule activities, as applicable

4. Contractor WBS to program WBS crosswalk

5. Schedule custom fields and activity codes dictionary and LOE field identification

6. Activity codes used to organize and filter the activities into categories as necessary to confirm a complete scope of work

7. Plans and documentation used for defining activities, such as the systems engineering plan, software development plan, risk management plan, and master test plan.

Likely Effects If Criteria Are Not Fully Met

1. If activities are missing from the schedule, then other best practices will not be met. If all activities are not accounted for, it is uncertain whether all activities are scheduled in the correct order, resources are properly allocated, missing activities will appear on the critical path, or a schedule risk analysis can account for all risk.

2. Failing to include all work for all deliverables, regardless of whether the deliverables are the responsibility of the government or contractor, can lead to program members' incomplete understanding of the plan and its progress toward a successful conclusion.

3. If the schedule does not fully and accurately reflect the program, it will not be an appropriate basis for analyzing or measuring technical work accomplished and may result in unreliable completion dates, time extension requests, and delays.

4. If government work is not captured in the IMS, the program manager will be less able to plan all the work and minimize the risk of government-caused delays.

5. Because the schedule is used for coordination, missing elements will hinder coordination efforts, increasing the likelihood of disruption and delays.

6. If the schedule is not planned in sufficient detail, then opportunities for process improvement (for example, identifying redundant activities), what-if analysis, and risk mitigation will be missed.

7. A schedule that does not emanate from a single start milestone activity and terminate at a single finish milestone activity is not properly constructed and may produce an erroneous critical path.

8. LOE activities can interfere with the critical path unless they are clearly marked and represented as summary or hammock activities designed for the purpose.

9. Too many milestones in the schedule can mask the activities necessary to achieve key milestones and can prevent the proper recording of progress.

10. Schedules that are defined at too high a level may disguise risk that is inherent in lower-level activities. Conversely, schedules that have too much detail make it difficult to manage progress.

11. Unless the schedule is aligned to the program WBS, management cannot ensure that the total scope of work is accounted for within the schedule.

12. Repetitive naming of activities makes communication difficult between teams, particularly between team members who are responsible for updating and integrating multiple schedules.

BEST PRACTICE 2: SEQUENCING ALL ACTIVITIES

Key Questions

1. Have the activities and logical relationships been determined by those executing the program?
2. Are the majority of the relationships within the detailed schedules finish-to-start?
3. Are predecessor links (with the exception of the start milestone) or successor links (with the exception of the finish milestone) missing?
4. Are any predecessors or successors dangling?
   a. Does each activity (except the start milestone) have an F–S or S–S predecessor that drives its start date?
   b. Does each activity (except the finish milestone and deliverables that leave the project without subsequent effect on the project) have an F–S or F–F successor that it drives?
5. Do summary activities have predecessor or successor links?
6. Do activities have start-to-finish links?
7. How much convergence (that is, several parallel activities converging at one major event) is there in the schedule? For activities that have many converging predecessors, do those predecessors have adequate float?
8. Does the schedule contain date constraints other than "as soon as possible"? Is each one justified in the schedule documentation?
9. Are lags or leads specified between the activities? Can these be more accurately characterized by improving logic or adding activity detail?

Key Documentation

1. Documentation justifies using hard and soft date constraints instead of activities' duration and logic.
2. Documentation justifies using lags and leads instead of activities' duration and logic.
3. Documentation justifies any activity that has no F–S or S–S predecessor or no F–S or F–F successor.

Likely Effects If Criteria Are Not Fully Met

1. The logical sequencing of events is directly related to float calculations and the critical path. If the schedule is missing dependencies or if activities are linked incorrectly, float estimates will be miscalculated. Incorrect float estimates may result in an invalid critical path and, thus, will not be reliable indicators of where resources can be shifted to support delayed critical activities.
2. That all interdependencies between activities are identified is necessary for the schedule to properly calculate dates and predict changes in the future. Without the right links, activities that slip early in the schedule do not transmit delays to activities that should depend on them. When this happens, the schedule will not allow a sufficient understanding of the program as a whole, and users of the schedule will

lack confidence in the dates and the critical path. Finally, when activities are not correctly linked, the program cannot use the IMS to identify disconnects or hidden opportunities and cannot otherwise promote efficiency and accuracy or control the program by comparing actual to planned progress.

3. Logical sequencing promotes a realistic workflow. If logic between activities is missing, program team members can misunderstand one another, especially regarding receivables and deliverables.

4. For scheduling software packages that include the option, summary activities should not have logic relationships because their start and finish dates are derived from lower-level activities. Summary logic hinders vertical traceability by obstructing the logic of lower-level activities.

5. A start-to-finish (S–F) link has the bizarre effect of directing a successor activity not to finish until its predecessor activity starts, in effect reversing the expected flow of sequence logic. The use of S–F logic is counterintuitive and overcomplicates schedule network logic.

6. The presence of "dangling activities" reduces the credibility of the calculated activity start and finish dates and the identity of the critical paths. The slip or elongation of an activity that has no logical successor will not reflect its effect on the scheduled start dates of successor activities.

    a. If an activity—other than the start milestone—does not have an F–S or S–S predecessor that drives its start date, the activity will start earlier if its duration is projected to be longer than originally believed. An earlier start may be illogical.

    b. If an activity—other than the finish milestone or deliverable that leaves the project—does not drive a successor by an F–S or F–F link, the implications of its running late or long are not passed on to any successor activity.

7. The ability of a schedule to forecast start and finish dates of activities and key events is directly related to the complexity and completeness of the schedule network. Unless complete network logic is established, the schedule cannot predict the effects on the program's planned finish date from, among other things, misallocated resources, delayed activities, external events, and unrealistic deadlines.

8. Because a logic relationship dictates the effect of an on-time, delayed, or accelerated activity on following activities, any missing logic relationship is potentially damaging to the entire network.

9. Path convergence issues can represent an unrealistic plan by implying that a large number of activities must be finished at the same time before a major event can occur as planned. An excess number of parallel relationships can indicate an overly aggressive or unrealistic schedule.

10. Hard date constraints that restrict activities to starting or finishing on a specific date must be justified by referring to some controlling event outside the schedule. Date constraints prevent activities from responding dynamically to network logic, including actual progress and availability of resources. They can seriously affect float calculations and the identification or continuity of the critical path and can mask both progress and delays in the schedule.

11. Hard and soft constraints interfere with the results of a schedule risk analysis because they prevent activity dates within the schedule from dynamically responding to changes in predecessor dates.

12. A customer-mandated date is not a legitimate reason to constrain an activity. A schedule is intended to be a dynamic, pro-active planning and risk mitigation tool that models the program and can be used to track progress toward important program milestones. Schedules with constrained dates can portray an artificial or unrealistic view of the program plan.

13. Constraints should be used only when necessary and only if their justification is documented because they override network logic and restrict how planned dates respond to accomplished effort or resource availability. The presence of a large number of activities with constraints is typically a substitute for logic and can mean that the schedule is not well planned and may not be feasible.

14. SNLT and FNLT constraints prevent activities from starting or finishing later than planned, essentially restricting the ability of any predecessor delays to affect their start and finish dates.

15. Applying constraints to represent the availability of resources requires constant manual upkeep of the schedule.

16. Mandatory start and finish constraints are the most rigid of all constraints because they do not allow the activity either to take advantage of time savings by predecessor activities or to slip in response to delayed predecessors or longer-than-scheduled durations.

17. The time to produce an external product should be represented by a reference or schedule visibility activity rather than a constrained milestone representing receipt of the product. By modeling vendor or contractor production as an activity, the program office can track the contractor's high-level progress and apply risk to the external production activity.

18. Lags must be justified because they may represent work or delay that may be variable while the lag is static. Lags should not be used to represent activities because they cannot be easily monitored or included in the risk assessment and do not take resources. Activities represented by lags are not, in fact, risk free.

19. Constantly updating lags manually defeats the purpose of a dynamic schedule and makes it particularly prone to error.

20. Using a lag with F–S logic is generally not good practice because it is generally not necessary. When it is, every effort should be made to break activities into smaller tasks and to identify realistic predecessors and successors so that logic interface points are clearly available for needed dependency assignments.

21. Leads are generally not valid. As negative lags, leads imply the unusual measurement of negative time and exact foresight about future events.

22. Using lags as buffers or margin for risk between two activities should be discouraged because the lags persist even as the actual intended margin is used up.

BEST PRACTICE 3: ASSIGNING RESOURCES TO ALL ACTIVITIES

Key Questions

1. What resources are specified and assigned to the activities? At what level of detail are resources specified (for example, as labor categories, organizations, or individual names)?
2. Are significant material and equipment resources described in the schedule?
3. Do summary activities or milestones have resource assignments?
4. How were resource estimates developed for each activity?
5. Has analysis ensured that resources are sufficient and available in each work period when needed?
   a. Is obtaining scarce resources to accomplish the work potentially difficult?
   b. Are more resources required than are available for some work periods? What is the plan for resolving resource deficiencies?
6. Has resource leveling been performed?
7. To what extent are the resource estimates in the schedule consistent with those in the program cost estimate?

Key Documentation

1. Basis of estimates for resource assumptions that align with resource estimates within the cost estimates.
2. A resource allocation planning document that defines resource profiles and tables for unique resources derived from the schedule.
3. Resource output from scheduling software across all project schedules.

Likely Effects If Criteria Are Not Fully Met

1. Information on resource needs and availability in each work period assists the program office in forecasting the likelihood that activities will be completed as scheduled. If the current schedule does not allow insight into the current or projected allocation of resources, then the risk of the program's slipping is significantly increased. Overallocated resources result in inefficiency (for example, staff are less productive because of extended overtime) or program delay from unavailable resources.
2. Resources must be considered in the creation of a schedule because their availability directly affects an activity's duration.
3. A schedule without resources implies an unlimited number of resources and their unlimited availability.
4. If there is no justification for allocating and assigning resources, the schedule will convey accuracy falsely.
5. Unrealistic peaks in forecasts of resource assignments represent the need for large amounts of resources near the end of work streams to finish deferred or delayed work on time. Often the quantity of resources and funding required at the peak is unrealistic.

6. If resource leveling causes enormous delays in the program finish date—for example, by many months or years—then the original resource assumptions, network logic, or activity durations must be examined for pragmatism.

7. Automatic resource leveling can lead to inefficient output by delaying activities if only partial resources are available and preventing activities from being partially accomplished while waiting for the full complement of resources to become available.

8. Incorrect resource assumptions (usually in the form of unwarranted optimism) will lend unreasonable credence to a resource-leveled schedule, and the resulting schedule will convey a false sense of precision and confidence to senior decision makers.

9. A schedule that has not reviewed and resolved resource use issues is not reliable.

10. If the baseline schedule does not identify the planned resources, it cannot be used to make important management decisions, such as reallocating resources from activities with significant float to critical activities that are behind schedule.

11. If the schedule does not have resource assignments, management's ability to monitor crew productivity, allocate idle resources, monitor resource-constrained activities, and level resources across activities is severely limited.

BEST PRACTICE 4: ESTABLISHING DURATIONS FOR ALL ACTIVITIES

Key Questions

1. Were durations determined from work to be done and realistic assumptions about available resources, productivity, normal interferences and distractions, and reliance on others?

2. For a detailed schedule, are durations short enough to be consistent with the needs of effective planning and program execution?

3. Are activities long in duration because of LOE or rolling wave planning?

4. Are LOE activity durations determined by the activities they support?

5. Did the person responsible for the activities estimate their durations?

6. Was the program duration determined by some target or mandated date?

7. Are durations based on appropriate calendars? Do any specific conditions necessitate special calendars, and are they addressed (for example, religious holidays, nonwork periods for climate, shift work, unavailability of resources)?

8. Are activity durations assigned inconsistent time units?

Key Documentation

1. How durations of work activities were estimated is documented at the appropriate level of detail. For instance, the basis of estimate includes the assumptions made to justify the durations assumed for the cost. These should be consistent with the durations at the same level of detail.

2. Documentation justifies nonstandard working calendars.

3. Documentation justifies excessively long durations, including the identification of LOE activities and how they were scheduled.

Likely Effects If Criteria Are Not Fully Met

1. If activities are too long, the schedule may not have enough detail for effective progress measurement and reporting.

2. If activities are too short, the schedule may be too detailed. This may lead to excessive work in maintaining the logic, updating the status of activities, and managing the many short-duration activities.

3. When durations are not based on the effort required to complete an activity, the resources available, resource efficiency, and other factors such as previous experience on similar activities, then there is little confidence in meeting the target deliverable date.

4. Schedules determined by imposed target completion dates rather than work and logic are often infeasible.

5. Durations estimated under optimal or "success-oriented" conditions will produce unrealistic program delivery dates and unreliable critical paths and could mask program risks.

6. Proper use of resource and task calendars usually precludes the need for soft constraints in schedules. But improperly defined task or resource calendars incorrectly represent the forecasted start, finish, and durations of planned activities. Ensuring realistic calendars provides for more accurate dates and may reveal opportunities to advance the work.

7. The default calendar in a schedule software package rarely has appropriate national holidays defined as exceptions and will not have specific blackout periods or other project-specific exceptions defined.

BEST PRACTICE 5: VERIFYING THAT THE SCHEDULE IS TRACEABLE HORIZONTALLY AND VERTICALLY

Key Questions

1. Is all logic in place and has the technical content of the schedule been validated?
2. Are major hand-offs and deliverables easily identified in the schedule? How are major hand-offs and deliverables negotiated and monitored?
3. Has horizontal traceability been demonstrated by observing the effects of delaying an activity by many days within the schedule or a similar shock to the network?
4. Are the key dates consistent between lower-level detailed working schedules and higher-level summary schedules? Do all lower-level activities roll up into higher WBS levels?
5. Do major milestones map between the schedule and management documents and presentations?

Key Documentation

1. All representations of the schedule are given as of a specific time. These may include different levels of the same schedule used in presentations as well as schedule representation using different platforms (scheduling or presentation packages) for different audiences.

2. The integration between summary, intermediate, and detailed schedules is demonstrated.

Likely Effects If Criteria Are Not Fully Met

1. If the schedule is not horizontally traceable, there may be little confidence in the calculated dates or critical paths.

2. Unless the schedule is horizontally traceable, activities whose durations are greatly extended will have no effect on key milestones.

3. Schedules that are not horizontally integrated may not depict relationships between different program elements and product hand-offs. When this happens, hand-offs of project subcomponents cannot be fully traced to the end product, leading to less effective program management.

4. Vertical traceability provides assurance that the representation of the schedule to different audiences is consistent and accurate. Without vertical traceability, there may be little confidence that all consumers of the schedule are getting the same correct schedule information.

5. Any logic errors between summary, intermediate, and detailed schedules will cause inconsistent dates between schedules and will cause different expectations between management and activity owners.

6. Unless the schedule is vertically traceable, lower-level schedules will not be consistent with upper-level schedule milestones, affecting the integrity of the entire schedule and the ability of different teams to work to the same schedule expectations.

BEST PRACTICE 6: CONFIRMING THAT THE CRITICAL PATH IS VALID

Key Questions

1. Is the critical path, or longest path (in the presence of date constraints), calculated by the scheduling software valid?
    a. Are any activities in the schedule missing logic or constrained without justification? Are these issues resulting in an unreliable critical path?
    b. Is the critical path a continuous path from the status date to the major completion milestones?
    c. Does the critical path start with a constraint so that other activities are unimportant in driving the milestone date? If so, is there justification for that constraint?
    d. Does the critical path include LOE activities? Is the critical path driven by activities of unusually long duration that are not considered planning packages?
    e. Is the critical path driven in any way by lags or leads?

2. Does management use the critical path to focus on activities that will detrimentally affect key program milestones and deliveries if they slip?

Key Documentation

1. Important program deliverables or milestones for which critical paths should be established are identified.
2. Printouts of the logic diagram indicate the longest paths to the important milestones, as well as critical paths based on total float to all major milestones.
3. Near-critical paths are identified.

Likely Effects If Criteria Are Not Fully Met

1. Without a valid critical path, management cannot focus on activities that will detrimentally affect the key program milestones and deliveries if they slip.
2. Unless the schedule can produce a true critical path, the program office will not be able to provide reliable timeline estimates or identify when problems or changes may occur and their effects on downstream work.
3. Successfully identifying the critical path relies on capturing all activities (Best Practice 1), properly sequencing activities (Best Practice 2), horizontal traceability (Best Practice 5), the reasonableness of float (Best Practice 7), accurate status updates (Best Practice 9), and—if there are resource limitations—assigning resources (Best Practice 3).
4. Unless the schedule is fully horizontally traceable, the effects of slipped activities on successor activities cannot be determined. If the schedule is missing dependencies or if activities are not linked correctly, float estimates will be miscalculated. Incorrect float estimates will result in an invalid critical path and will hinder management's ability to allocate resources from noncritical activities to those that must be completed on time.
5. LOE activities should not drive the schedule. If LOE is critical, management has no indication of which activities can slip and which will respond positively to additional resources to reduce the risk of finishing late.
6. The review and analysis of near-critical paths is important because their activities are likely to overtake the existing critical path and drive the schedule.

BEST PRACTICE 7: ENSURING REASONABLE TOTAL FLOAT

Key Questions

1. Are the total float values that the scheduling software calculates reasonable and do they accurately reflect true schedule flexibility?
2. Are excessive values of total float being driven by activities that are missing logic?
3. Is total float monitored? Does management have a plan to mitigate negative total float?
4. Does management rely on free float to level resources or reassign resources to assist critical activities?

Key Documentation

The program team can use a list of activities sorted by their total float values to determine whether the total float values correctly reflect flexibility in the program schedule.

Likely Effects If Criteria Are Not Fully Met

1. If the schedule is missing activities or dependencies or if it links activities incorrectly, float estimates will not be accurate. Incorrect float estimates may result in an invalid critical path and an inaccurate assessment of program completion dates. In addition, inaccurate values of total float falsely depict true program status, which could lead to decisions that may jeopardize the program. For example, if activities are not linked correctly to successors, total float will be greater than it should be.

2. Without accurate values of total float, it cannot be used to identify activities that could be permitted to slip and thus release and reallocate resources to activities that require more resources to be completed on time.

3. Negative float indicates that not enough time has been scheduled for the activity and is usually caused by activities taking longer or starting later than planned, making target dates infeasible. The program may have to take some corrective action or the negative float may act as a threat to the program end date.

4. Too little float built into the schedule may indicate insufficient time to recover from delay without the program's completion date slipping.

BEST PRACTICE 8: CONDUCTING A SCHEDULE RISK ANALYSIS

Key Questions

1. Was an SRA performed to determine the confidence level in achieving the program schedule and other key dates?
    a. Was the schedule checked to ensure that it meets best practices before the simulation was conducted?
    b. Are there data fields within the schedule for risk analysis such as optimistic, most likely, and pessimistic durations?
    c. Were uncertainties in activity durations statistically correlated to one another?
    d. How much schedule contingency was selected and what is the probability of meeting the completion date?
    e. Did the SRA identify activities during the simulation that most often ended up on the critical path, so that near-critical path activities can be closely monitored?
2. Was a risk register used as an input to schedule development?
    a. Was the risk register used in identifying the risk factors potentially driving the schedule before the SRA was conducted?
    b. Once the SRA was conducted, were risks prioritized by probability and magnitude of effect?
3. Are the SRA data, assumptions, and methodology available and documented?
4. Are the probabilities and impact ranges reasonable and based on information gathered from knowledgeable sources? Is there evidence of bias in the risk data?
5. How is the use of schedule contingency controlled and authorized?
6. Is an SRA performed periodically to reflect actual progress and changes in risks?

Key Documentation

1. A risk register with prioritized risks.
2. SRA documentation that includes assumptions, methodology, data, data normalization techniques, and findings.
3. A listing of people interviewed or included in risk interviews along with their organizations, positions, and expertise.
4. The schedule risk analysis file.

Likely Effects If Criteria Are Not Fully Met

1. If a schedule risk analysis is not conducted, the following cannot be determined:
    a. the likelihood of the program's completion date,
    b. how much schedule risk contingency is needed to provide an acceptable level of certainty for completion by a specific date,
    c. risks most likely to delay the program,
    d. the paths or activities that are most likely to delay the program.
2. Because activity durations are uncertain, the identity of the true critical path is unknown unless a schedule risk analysis has been performed. An SRA can identify the paths that are most likely to become critical as the program progresses so that risk mitigation can lessen the effect of any delays.
3. Unless a statistical simulation is run, calculating the completion date from schedule logic and the most likely duration distributions will tend to underestimate the program's overall critical path duration.
4. If the schedule risk analysis is to be valid, the program's schedule must reflect reliable logic and clearly identify the critical path. If the schedule does not follow best practices, confidence in the SRA results will be lacking.
5. If the program does not have sufficient schedule reserve, then risk mitigation actions and schedule issues from unforeseen events may not be managed without a schedule delay.
6. If the task durations are not correlated to one another, the uncertainty on the critical path duration may be underestimated.

BEST PRACTICE 9: UPDATING THE SCHEDULE USING LOGIC AND PROGRESS

Key Questions

1. Is progress recorded regularly? Has the schedule been updated recently as planned? Is the status date recorded?
2. Is at least one in-progress activity critical?
3. Do any activities have start or finish dates in the past without actual start or finish dates? Do any activities have actual start or finish dates in the future?
4. Is responsibility for changing or statusing the schedule assigned to someone who has the proper training and experience in CPM scheduling?

5. Were any activities started or completed out of sequence? If so, was the logic retained, or did the scheduler use progress override?

6. Does a schedule narrative accompany each status update and include the following?
   a. the status of key milestone dates, including the program finish date;
   b. the status of key hand-offs or giver/receiver dates;
   c. explanations for any changes in key dates;
   d. changes in network logic, including lags, date constraints, and relationship logic and their effect on the schedule time;
   e. a description of the critical paths, near-critical paths, and longest paths along with a comparison to the previous period's paths; and
   f. a description of any significant scheduling software options that changed between update periods, such as the criticality threshold for total float, progress override versus retained logic and whether resource assignments are progressed along with duration.

7. Is the schedule structure examined after each update to ensure that no logic is missing, constraints are necessary, and no activities impede the ability of the schedule to dynamically forecast dates?

Key Documentation

1. The schedule narrative.
2. The schedule shows actual and planned dates, remaining duration for in-process activities, and the status date.
3. Copies of program management review (PMR) briefings are available to verify whether schedule status is discussed and consistent with the schedule.

Likely Effects If Criteria Are Not Fully Met

1. If the schedule is not continually monitored to determine when forecasted completion dates differ from planned dates, then it cannot be used to determine whether schedule variances will affect downstream work.
2. Maintaining the integrity of the schedule logic is not only necessary to reflect true status but also required before conducting a schedule risk analysis. If the schedule has not been updated, then it is impossible to tell what activities have been completed, are in progress, are late, and are planned to start on time.
3. A schedule that has not been updated will not reflect what is actually occurring on the program and hence may have inaccurate completion dates and critical paths. When this is the case, management cannot use the schedule to monitor progress and make decisions regarding risk mitigation, resource allocations, and so on.
4. Unless a status date is provided, the schedule cannot be used to reliably convey effort spent and remaining.
5. An out-of-sequence activity causes degradation of the schedule and requires addressing. A schedule with progress remaining out of sequence may have the wrong logic in place and, hence, inaccurate critical paths and completion dates.
6. If unfinished work remains in the past, the schedule no longer represents a realis-

tic plan to complete the program, and team members will lose confidence in the model.

7. At least one in-progress activity is critical. If not, it is most likely that date constraints or external dependencies are separating successor activities from in-progress activities. Such breaks in the critical or longest path represent weak or incomplete logic, causing a lack of credibility in the identity of the path and the schedule dates.

8. Without a documented, consistently applied schedule change control process, program staff might continually revise the schedule to match performance, hindering the program manager's insight into the true performance of the program. Good documentation helps with analyzing changes in the program schedule and identifying the reasons for variances between estimates and actual results, thereby contributing to the collection of cost, schedule, and technical data that can be used to support future estimates.

9. Unless the schedule is kept updated, trend reports and analyses that highlight problems will not be useful in mitigating future delays.

10. Unless progress records are archived, historical data necessary for resource, work, and productivity assumptions for future analogous programs will not be available. If sufficient attention is paid to recording the way work is performed, the resulting archived data will help improve the accuracy and quality control of future similar programs.

BEST PRACTICE 10: MAINTAINING A BASELINE SCHEDULE

Key Questions

1. Is the baseline schedule the basis for measuring performance?
2. Does a schedule basis document exist? Does the document
    a. describe the general approach to the program?
    b. describe the overall structure of the IMS, including the scope and purpose of projects, staff responsible for each project, the relationship between projects, a WBS dictionary, the status delivery dates for each project, and a list of key hand-off products and their estimated dates?
    c. describe the settings for key options for the scheduling software?
    d. provide an overview of the assumptions and ground rules, including justification for calendars and any lags, constraints, or long activity durations?
    e. Provide an appropriately detailed rationale for the basic approach to estimating key activity durations and justification of the estimating relationship between duration, effort, and assigned resource units?
    f. contain a dictionary of abbreviations, acronyms, and custom fields?
    g. describe the use of resources within the schedule?
    h. describe the critical risks prioritized in a schedule risk analysis as well as schedule contingency?
    i. discuss the derivation of the critical paths and longest path and justify excessive total float?

3. Are changes to the baseline schedule reviewed and approved according to the schedule change control process?
4. Is trend analysis performed, such as monitoring start and finish dates, available float, and available schedule contingency?

Key Documentation

1. The designated baseline schedule.
2. A description of the schedule change control process.
3. The current schedule change control log.
4. The schedule basis document.

Likely Effects If Criteria Are Not Fully Met

1. Without a formally established baseline schedule to measure performance against, management cannot identify or mitigate the effect of unfavorable performance.
2. Good documentation helps with analyzing changes in the program schedule and identifying the reasons for variances between estimates and results, thereby contributing to the collection of cost, schedule, and technical data that can be used to support future estimates.
3. Thorough documentation is essential for validating and defending a baseline schedule. A well-documented schedule can convincingly argue for a schedule's validity and can help answer decision makers' and oversight groups' probing questions. A well-documented schedule is essential if an effective independent review is to ensure that it is reliable.
4. If changes are not controlled and fully documented, performance cannot be accurately measured against the original plan. Undocumented or unapproved changes will hamper performance measurement and may result in inaccurate variance reporting, inconsistent stakeholder versions of the plan, and unreliable schedule data.
5. Without a schedule change control process, traceability for all status updates will be unreliable, and there will be no guarantee that stakeholders are using the same version of the schedule.
6. Unless schedule variances are monitored, management will not be able to reliably determine whether forecasted completion dates differ from the planned dates.
7. Without trend analysis, management will lack valuable information about how a program is performing. Knowing what has caused problems in the past can help determine whether they will continue in the future.

APPENDIX III

SCHEDULING AND EARNED VALUE MANAGEMENT

As we note throughout the schedule guide, the success of a program depends in part on having an integrated and reliable master schedule that defines when and how long work will occur and how each activity is related to the others. The scheduling best practices demonstrated in this guide apply to all schedules, but additional considerations apply when scheduling in an earned value management (EVM) environment. This appendix briefly discusses these considerations and the associated EVM terms.

Chapters 18–20 of the GAO Cost Guide address the details of EVM, a project management tool that integrates the technical scope of work with schedule and cost elements for planning and control.[44] EVM is designed to integrate cost estimation, schedule development, system development oversight, and risk management. It compares the value of work accomplished in a given period with the value of the work planned for that period. It serves as a means of analyzing cost and schedule performance. By knowing what the planned cost is at any time and comparing that value to the planned cost of completed work and to the actual cost incurred, analysts can measure a program's cost and schedule status.

Without knowing the planned cost of completed work and work in progress (that is, earned value), management cannot determine true program status. Earned value provides the information necessary for understanding the health of a program and provides an objective view of program status. Moreover, because EVM provides data in consistent units (usually dollars or labor hours), the progress of vastly different work efforts can be combined. For example, earned value can be used to combine feet of cabling, square feet of sheet metal, or tons of rebar with effort for systems design and development. That is, earned value can be employed as long as a program is broken down into well-defined and objectively measured tasks.

Federal Acquisition Regulation (FAR) Subpart 34.2 requires an EVM system for major acquisitions for development. The government may also require the use of EVM for other acquisitions, in accordance with agency procedures. When a program operates within an EVM environment, the EVM system should meet the intent of the 32 guidelines from American National Standards Institute/Electronics Industries Alliance ANSI/EIA-748, a national standard for EVM systems.

---

[44]See Cost Guide, GAO-09-3SP.

Below, for scheduling best practices that are affected, we discuss the applicable EVM guidelines and specific EVM practices for schedules used within this type of environment. Specifically, we address

- work scope covered under EVM (Best Practice 1 and Guideline 1),
- establishing budgets and developing a performance measurement baseline (Best Practice 3 and Guidelines 3 and 8–15),
- control accounts (Best Practice 3 and Guideline 5),
- updating durations and setting objective measures to claim earned value (Best Practice 9 and Guidelines 7 and 22), and
- maintaining a baseline schedule and change control processes (Best Practice 10 and Guidelines 28–32)

BEST PRACTICE 1 – CAPTURING ALL ACTIVITIES

EVM Guideline 1 states that authorized program work elements should be defined. A WBS is usually used to do this. It is a representation of the work scope and breaks down all authorized work into the proper elements. Work is authorized at the control account level through a work authorization process. Work authorization documents define the work to be accomplished and include task outputs, deliverables, an associated budget, and authorizing signatures.

While not explicitly stated in the guidelines, some organizations interpret authorized work as work that has been placed under contract. Some auditing agencies, such as DOD Defense Contract Management Agency (DCMA), restrict their assessment of whether a schedule includes all effort to only ensuring that the scope of work that is under contract is included in the contractor's schedule. They will not assess scope that may be necessary to complete the program but is not under contract.[45] In Best Practice 1, we state that the IMS should reflect all effort necessary to successfully complete the program and that it is a comprehensive plan of all government, contractor, subcontractor, and key vendor work. The government program manager is responsible for managing the scope of all project work, not just the scope placed on contract.

However, a contractor project schedule, as a subset of the overall government program effort, will include only contractually authorized work because contractors are obligated to plan activities required by, and limited to, the contract. Thus, when reviewing schedules that support an EVM baseline, only the work that is under contract will be included. In some instances in which the government program manager monitors government effort with EVM, EVM guidelines will apply to all effort used to define the PMB. EVM guidelines are applicable to contracted effort and possibly to government effort as well, depending on the agency and the level of responsibility to which EVM is assigned. The auditor should keep these considerations in mind when assessing the schedule.

---

[45]DCMA's schedule assessment process is explained in detail in appendix VII.

BEST PRACTICE 3 – ASSIGNING RESOURCES TO ALL ACTIVITIES

We discuss fully loading the schedule with resources to provide the basis for the PMB in Best Practice 3. When a schedule is fully resource loaded, budgets for direct labor, travel, facilities, equipment, material, and the like are assigned to both work and planning packages so that total costs to complete the program are identified at the outset. Additionally, Best Practice 10 describes the importance of maintaining a baseline schedule. However, managers should consider additional best practices when developing a PMB because it forms the fundamental basis for measuring progress using EVM.

EVM Guidelines 3 and 8 discuss the actions to establish the PMB. It is important to establish and maintain a valid schedule baseline to ensure that EVM data being reported are reliable. Therefore, the entire schedule must be baselined because the IMS is the source of time-phasing for all control accounts and work packages that make up the project's PMB. Similar to Best Practice 1, this requirement applies to contracted effort and may apply to the government effort as well, depending on the agency and the level to which EVM is assigned. For efforts under contract, the baselines for cost and schedules are required to support the integrated baseline review (IBR). In reality, to successfully prepare for the IBR, baselines must be set (either formally or informally) much earlier than the IBR deadline to ensure credibility. More detail regarding PMB development and IBRs is in Chapter 18 of the GAO Cost Guide.

In building the PMB, fully resource loading the schedule is the easiest way to adequately time-phase costs for performance measurement. If the schedule is not fully loaded, then determining how costs are phased over time will be much more complex and difficult and the auditor may not be able to trace the logic.

With regard to control accounts, Best Practice 1 states that the WBS should progressively deconstruct the deliverables of the entire effort through lower-level elements. In EVM Guideline 5, the control account is created at the intersection of the WBS and the organizational breakdown structure (OBS), and it is at this level that actual costs are collected and variances from the baseline plan are reported. The control account is the focal point for work authorization and performance measurement. Because several organizations can be working on the same WBS element, each WBS element may have multiple control accounts. Each control account has staff who are assigned responsibility for managing and completing the work.

Below the control account level, the effort is further broken down into work packages and planning packages. It is at the work package level that detail activities can be identified in a schedule. As we discuss in Best Practice 3, a schedule incorporates different levels of detail, depending on the information available. Under EVM, work packages and planning packages are assigned to control accounts. Detailed activities represent work packages that are typically near-term effort. Ideally, work packages are typically 4 to 6 weeks long and require specific effort to meet control account objectives. They are defined by the persons who authorize the effort and how the work will be measured and tracked.

Work packages may be represented by a single detail activity or they may be broken down into more well-defined, lower-level tasks. Effort beyond the near-term that is less well defined may be represented as planning packages. Planning packages are used specifically as budget holding accounts for future work within a control account that cannot be planned in detail. The planning package, while not planned in detail, is sequenced logically within the schedule network, associated with a WBS element, and assigned resources.

Another representation of effort may also reside in summary level planning packages (SLPP) which are not assigned to control accounts. SLPPs are temporary and identify scope, schedule, and associated budget that cannot be practically assigned to a control account. SLPPs should be assigned to control accounts at the earliest opportunity. Therefore, the auditor should be mindful that only near-term work packages will be detail planned and should represent short duration of discrete efforts. Planning packages and SLPPs are likely to be much longer in duration and will be detail planned as part of the rolling wave process.

BEST PRACTICE 9 – UPDATING THE SCHEDULE USING ACTUAL PROGRESS AND LOGIC

In Best Practice 9, we discuss statusing the schedule and how it generally takes the form of updating either activity durations or work. Updating an activity's actual and remaining duration is the most common method of recording progress because it is the easiest to do; however, duration updates can be easily misconstrued. Because an update is in terms of activity duration, it denotes the passage of time from the start date, not the amount of work performed. Some program managers may wish to update work by resources to track actual effort. This practice takes more time than updating duration but provides better forecasts of remaining effort.

EVM Guidelines 7 and 22 address statusing the schedule and measuring performance. Under EVM, updating activity status requires identifying objective measures for use in statusing activity progress. Using objective measures, rather than updating durations, allows for measuring work accomplished and permits an accurate comparison to the work planned. These measures allow for developing variances that provide visibility into project performance that helps the project manager properly focus attention on areas requiring improvement. The three types of measure are discrete effort, level of effort, and apportioned effort. Discrete effort measures are used with tasks that can be directly measured and are related to the completion of specific end products. Some discrete effort methods are described in table 8.

Table 8: Seven Measures of Effort

Method	Description
0/100	No performance is taken until a task is finished
50/50, 25/75, etc.	Half the earned value is taken when the task starts, the other half when it is finished; other percentage combinations can be used
Percentage complete	Performance is measured by percentage of work completed. This should be based on underlying, quantifiable measures as much as possible (e.g., number of drawings completed) and can be measured by the statusing of the resource loaded schedule. The percentage complete for each work package is the cumulative value of the work accomplished to date divided by the total value of the work package
Weighted milestone	Performance is taken as defined milestones are accomplished; objective milestones (weighted by importance) are established monthly, and the budget is divided by milestone weights; as milestones are completed, value is earned
Units complete	Performance is measured by counts of similar products. This method is used in construction and manufacturing
Equivalent units	Performance is taken for completed units or the fractional equivalent of completed full units. This method is used in construction and manufacturing

Source: DOD | PMI | GAO-16-89G.

As described in Best Practice 1, level-of-effort activities are related to the passage of time and have no physical products or defined deliverables. One example is program management.

Apportioned effort is effort that by itself is not readily divisible into short-span work packages but is related in direct proportion to an activity or activities with discrete measured effort. Apportioned effort work packages can be defined as discrete work packages, but apportioned effort tasks are unique because they are closely dependent on another discrete work package. Examples include quality control (QC) responsibilities associated with pipefitting or pouring concrete. These QC activities should be hammocked with their related activities, and their earned value should be based on the related activities' earned value.

Additional discussion of objective measures for earned value is in Chapter 18 of the GAO Cost Guide. The auditor should be able to clearly understand how earned value is being calculated and whether the method used is appropriate for adequate statusing of the true effort performed.

BEST PRACTICE 10 – MAINTAINING A BASELINE SCHEDULE

Under Best Practice 3, we discuss the importance of developing the PMB and establishing and maintaining a valid baseline schedule to ensure that EVM data are reliable. In Best Practice 10, we discuss the importance of maintaining a baseline schedule and the change control process that governs when and how technical and programmatic changes are applied to the baseline. We further say that the extent to which the IMS is subject to a change control process is decided by program management and depends on project size, scope, risk, and complexity. When EVM is required, the change control process is more restrictive, and the program or its contractor should have an approved change control process for PMB changes from authorized changes and replanning.

Authorized changes must be punctually recorded in the EVM system and incorporated into planning. Controlling documents must be updated before new work is started. EVM guidelines 28–32 address documenting and controlling changes to the PMB. The guidelines specifically discuss incorporating changes in a timely manner, reconciling current to prior budgets, controlling retroactive changes, preventing unauthorized revisions, and documenting PMB changes. The auditor should be able to identify the baseline schedule as well as the most recently statused schedule in order to verify that variances reported in the EVM system are valid.

APPENDIX IV

THE FORWARD AND BACKWARD PASS

Early and late activity dates are determined by the logical sequence of effort that planners lay out. Network logic calculates activity dates that define both when an activity may start and finish and when an activity must start and finish to meet a specified program completion date.

Suppose house construction and exterior finishing have been completed, and the necessary electrical, gas, and water inspections are complete. Several activities remain before the owner can occupy the house—specifically, utility systems must be started up and tested. Workers must

- set the electricity meter,
- start up and test the electrical system,
- set the gas meter,
- start up and test the furnace and air conditioner,
- set the water meter, and
- start up and test the plumbing fixtures.

Once these activities are completed, then the start-up and testing phase is completed and the owner can occupy the house. These activities must happen in a specified order:

- The electricity, gas, and water meters cannot be set until the inspections are completed.
- The electrical system cannot be tested until the meter is set.
- The furnace and air conditioner cannot be tested until the gas and electricity meters are set.
- The plumbing fixtures cannot be tested until the gas and water meters are set.
- The start-up and test phase is complete when the electrical system, furnace, air conditioner, and plumbing fixtures are tested.

Table 9 shows the expected durations of each activity, given the estimated resources.

Table 9: Expected Durations and Estimated Resources in House Construction

Activity	Resource name	Resource units	Work hours	Duration in days
Set electricity meter	Electrician	2	16	1
	Electrical service technician	1	8	
Start up and test electrical system	Electrician	1	16	2
Set gas meter	Gas service technician	2	16	1
Start up and test furnace and air conditioner	HVAC technician	1	8	1
Set water meter	Water service technician	2	16	1
Start up and test plumbing fixtures	Plumber	2	32	2

Source: GAO | GAO-16-89G.

Assuming the inspections are completed by Monday, January 26, the electricity, gas, and water meters can be set on January 27. Then, assuming that all succeeding activities are related by finish-to-start relationships, the start-up and testing network will appear as in figure 51.

Figure 51: Start-Up and Testing Network

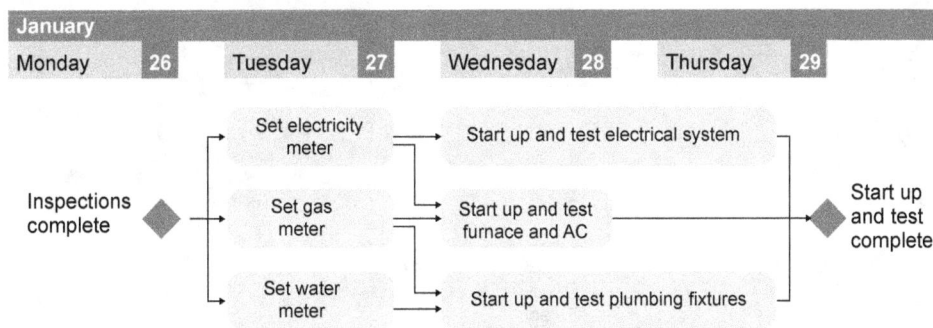

Source: GAO. | GAO-16-89G

The activities in figure 51 are planned according to forward scheduling; that is, activities begin as soon as possible according to their logic relationships. For example, once the electricity meter is set, starting up and testing the electrical system can begin. Calculating the earliest dates when activities can start and finish—given their predecessor and successor logic and durations—is known as the forward pass. In the forward pass, durations are added successively through the network.

If inspections complete at the end of the business day on Monday, January 26, then the work associated with setting the electricity, gas, and water meters can begin on Tuesday, January 27. This is the early start (ES) for these activities and is noted in a box in the upper left corner of the text box representing each activity in figure 52. To calculate the early finish (EF) of each activity, the duration is added to the ES. One day must be subtracted to account for the full day of work available between the early start and early finish. That is,

$$EF = ES + duration - 1$$

The EF is noted in the upper right corner of the activity's text box. The durations are noted between the ES and EF boxes. In these simple cases of 1-day activities, the early start and early finish dates are the same day. The numbers above the date boxes represent the cumulative duration of the project as work progresses along those dates.

Figure 52: Early Start and Early Finish Calculations

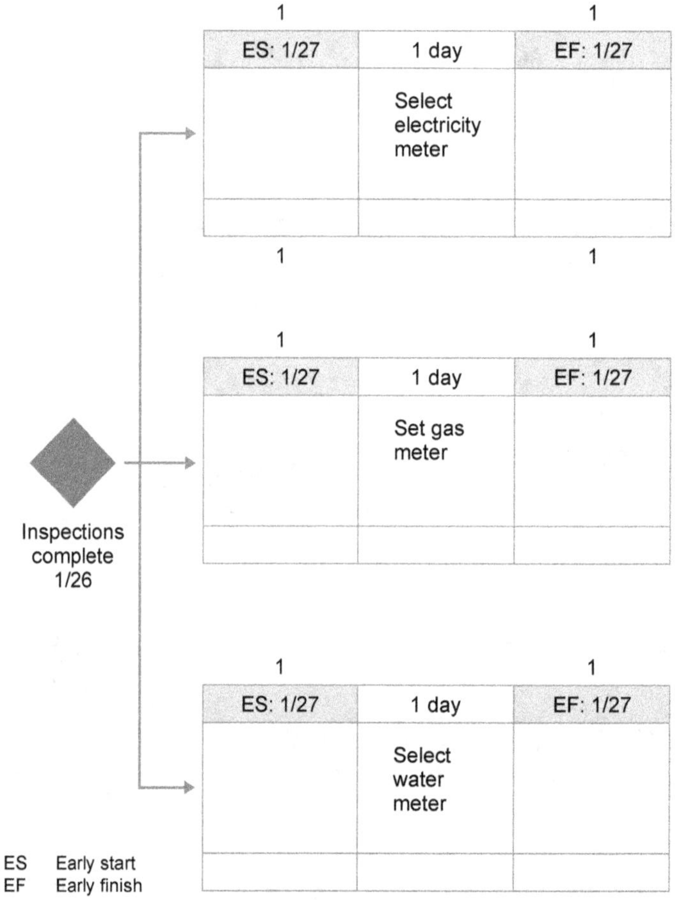

Source: GAO. | GAO-16-89G

Early starts of successor activities are calculated according to the logic of the network. For a single finish-to-start relationship, the early start of the successor activity is simply the predecessor's early finish plus 1 day. The "start up and test electrical system" activity can begin the day after "set electricity meter" finishes, as shown in figure 53. The EF for "start up and test electrical system" is its ES (January 28) plus its 2 days of duration (January 30) and minus 1 day (January 29).

Figure 53: Successive Early Start and Early Finish Calculations

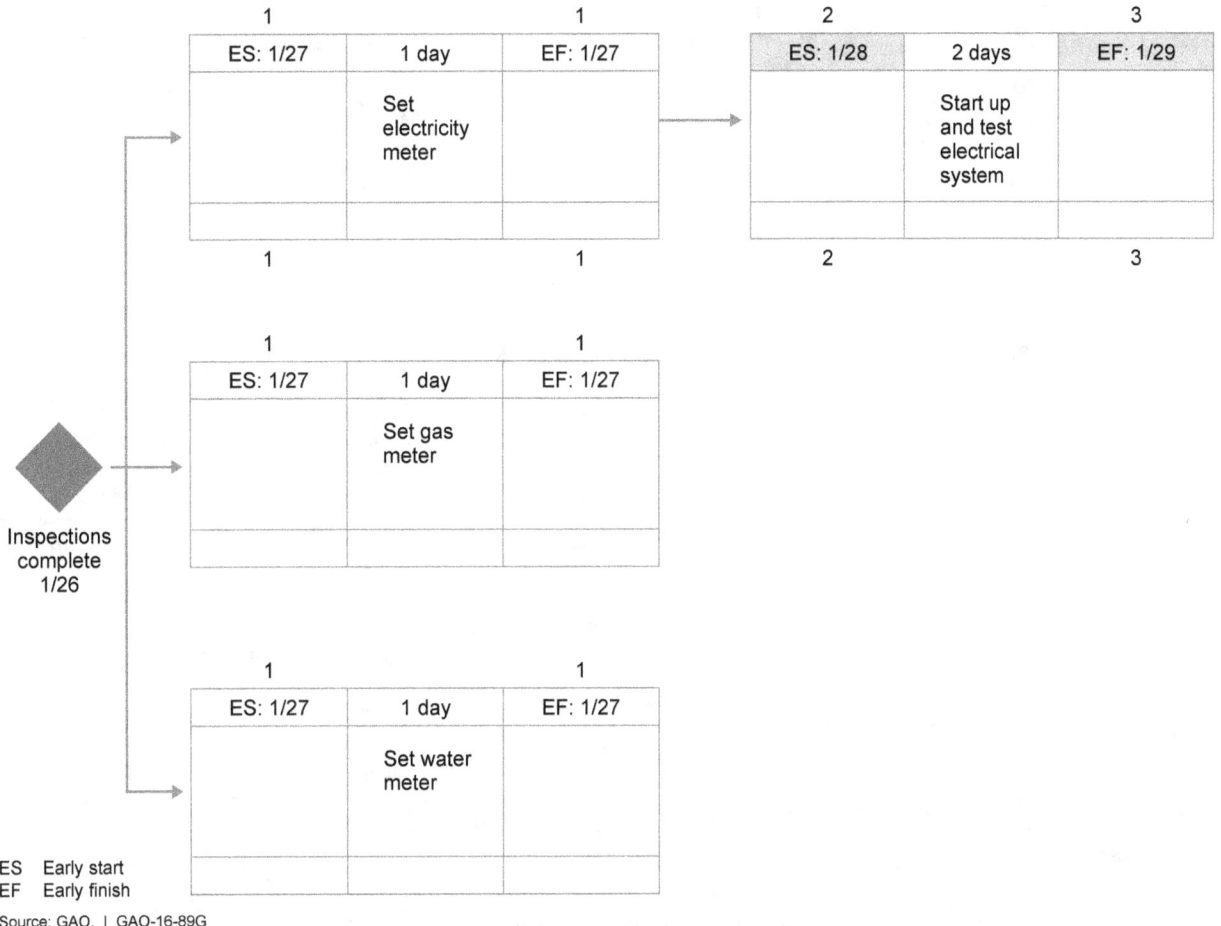

ES Early start
EF Early finish

Source: GAO. | GAO-16-89G

When an activity has two or more predecessor activities, the ES is calculated with the latest EF of its predecessor activities. That is, an activity cannot begin until the latest predecessor finishes. In figure 54, "start up and test furnace and AC" cannot begin until both the electricity meter and the gas meter are set. Setting the meters takes a day for each one, but they happen concurrently so they both finish on January 27. Therefore, starting up and testing the furnace and AC can start as early as January 28. Likewise, plumbing fixtures testing cannot begin until the gas meter and the water meter are set. But because these are both 1 day long and occur on the same day, testing plumbing fixtures can begin as early as January 28.

Figure 54: Complete Early Start and Early Finish Calculations

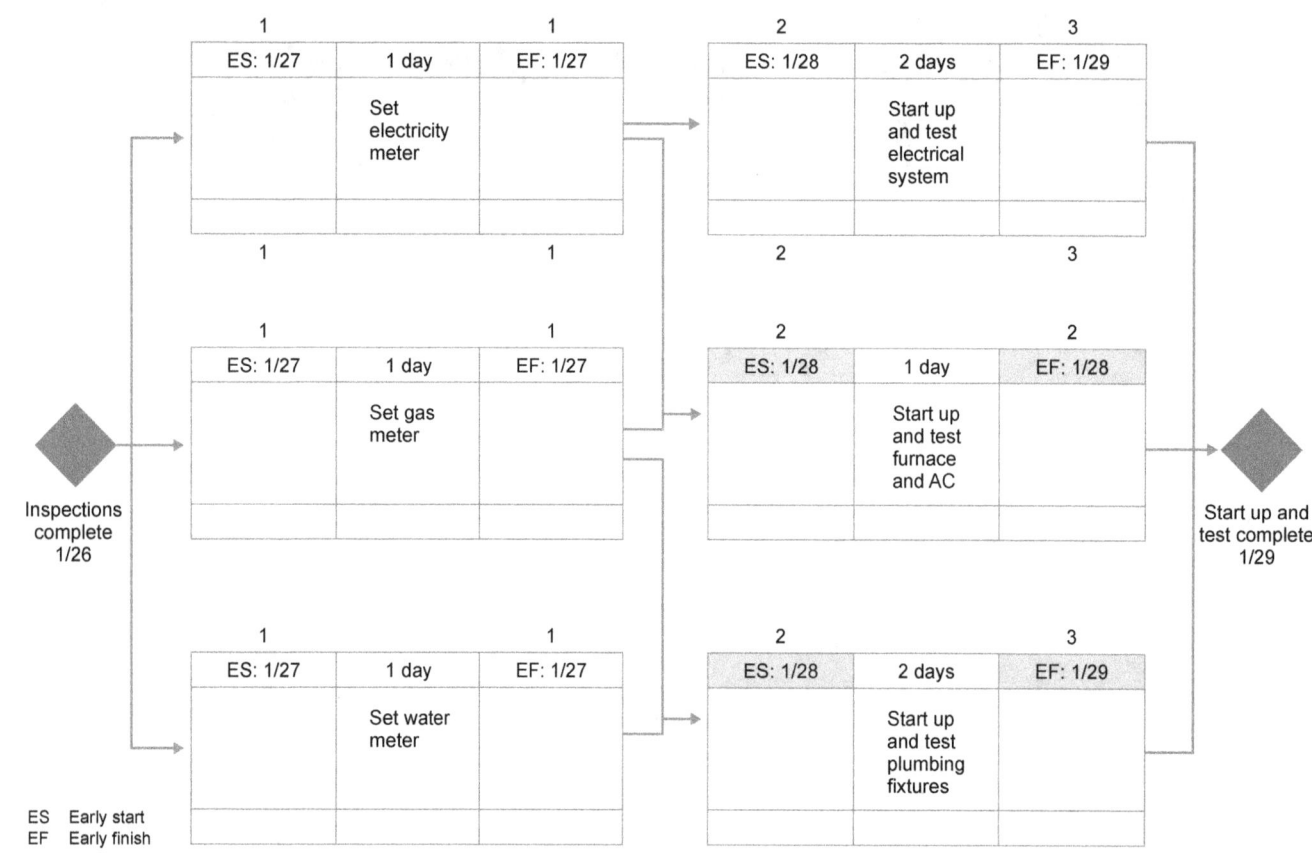

ES   Early start
EF   Early finish

Source: GAO. | GAO-16-89G

Note that testing plumbing fixtures is expected to take 2 days as well. Its EF is calculated as its ES (January 28) plus 2 days (January 30) minus 1 day (January 29). Finally, the ES of the "start up and test complete" milestone is calculated with the latest EF of its three predecessors. Both "start up and test plumbing fixtures" and "start up and test electrical system" have EF dates of January 29. Therefore, the earliest when "start up and test complete" could possibly occur is January 29 (milestones have no duration and therefore occur immediately after their latest predecessor).

As can be concluded by these calculations, the ES and EF dates are the earliest dates when an activity may start and finish. We are also interested in the latest dates when an activity must start or finish. That is, what are the latest dates when activities must start and finish to finish a project by a given date? These dates are calculated by the backward pass. Once the forward pass has been calculated, the backward pass determines the latest possible start and finish dates for activities.

The backward pass essentially calculates how long activities can wait to start or finish with the project still completing on time. The backward pass begins with the EF of the project. In the house construction set-up and test example, each of the three testing activities has a late finish (LF) of January 29. This is equal to the latest EF of the testing activities and, likewise, the date of the completion milestone. In figure 55, the LF is noted in the bottom right corner of each activity. To calculate an activity's late start (LS), the duration is subtracted from its LF, and 1 day is added to account for the full day of work between the two dates:

$$LS = LF - duration + 1$$

The LS is noted in the bottom left corner of the activity's text box in figure 55.

Figure 55: Late Start and Late Finish Calculations

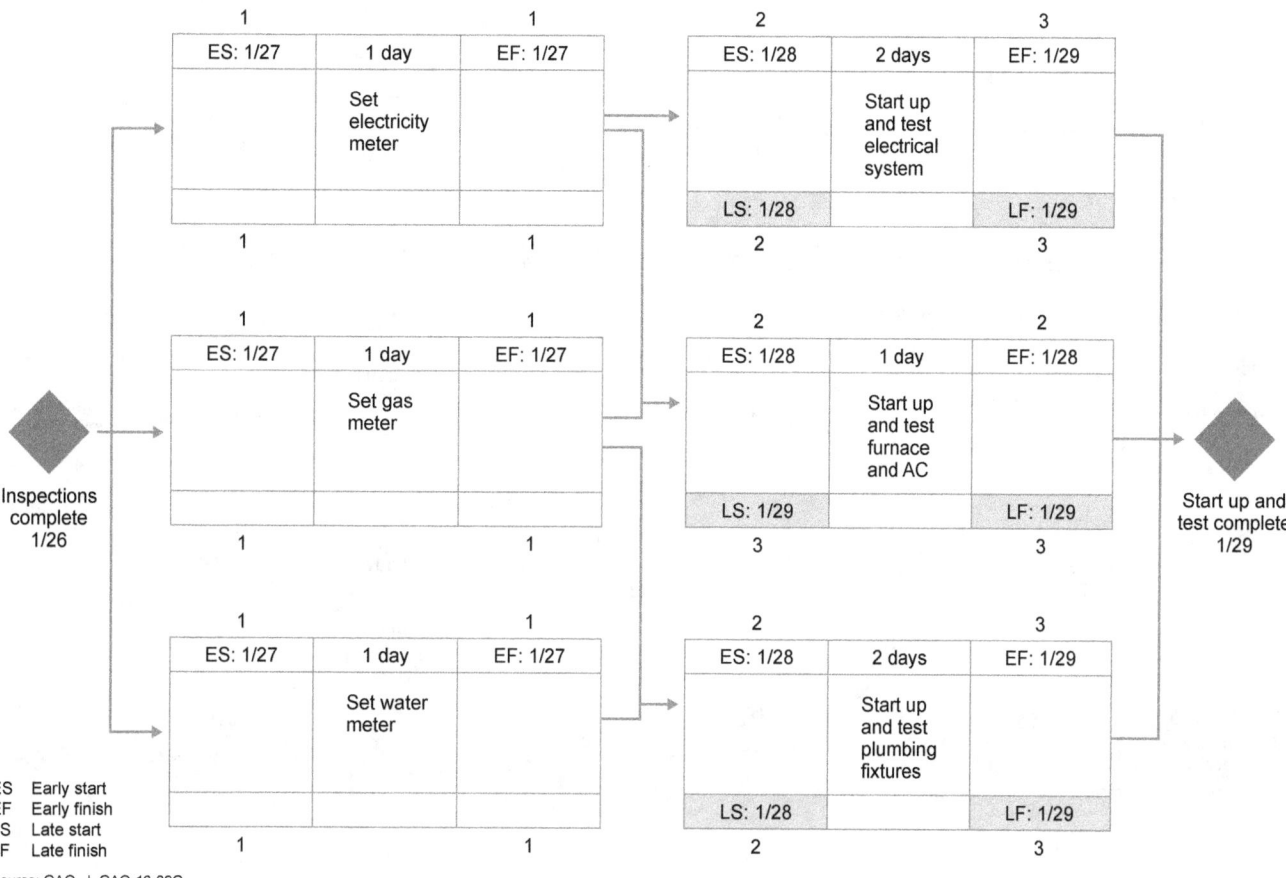

ES   Early start
EF   Early finish
LS   Late start
LF   Late finish

Source: GAO. | GAO-16-89G

In general, an activity's LF is equal to its successor's LS minus 1 day:

$$LF = LS - 1$$

When an activity has two or more successor activities, its LF is derived from the earliest LS of its successors. Stated another way, an activity does not need to finish until its successor must start; in the case of multiple successors, the earliest successor start date drives the activity's LF. In the following example, "set gas meter" has two successors: "start up and test furnace and AC" and "start up and test plumbing fixtures." The late start for testing plumbing fixtures is January 28, and the late start for testing the furnace and AC is January 29. The LF for "set gas meter" uses the earliest LS of its successors, which in this case is January 28 for "start up and test plumbing fixtures."

The successors for "set electricity meter"—"start up and test electrical system" and "start up and test furnace and AC"—also have different LS dates (January 28 and 29). Its LF is therefore January 27, 1 day before the earliest LS date. The complete forward pass and backward pass calculations are shown in figure 56.

Figure 56: Early and Late Dates of a Start-Up and Testing Network

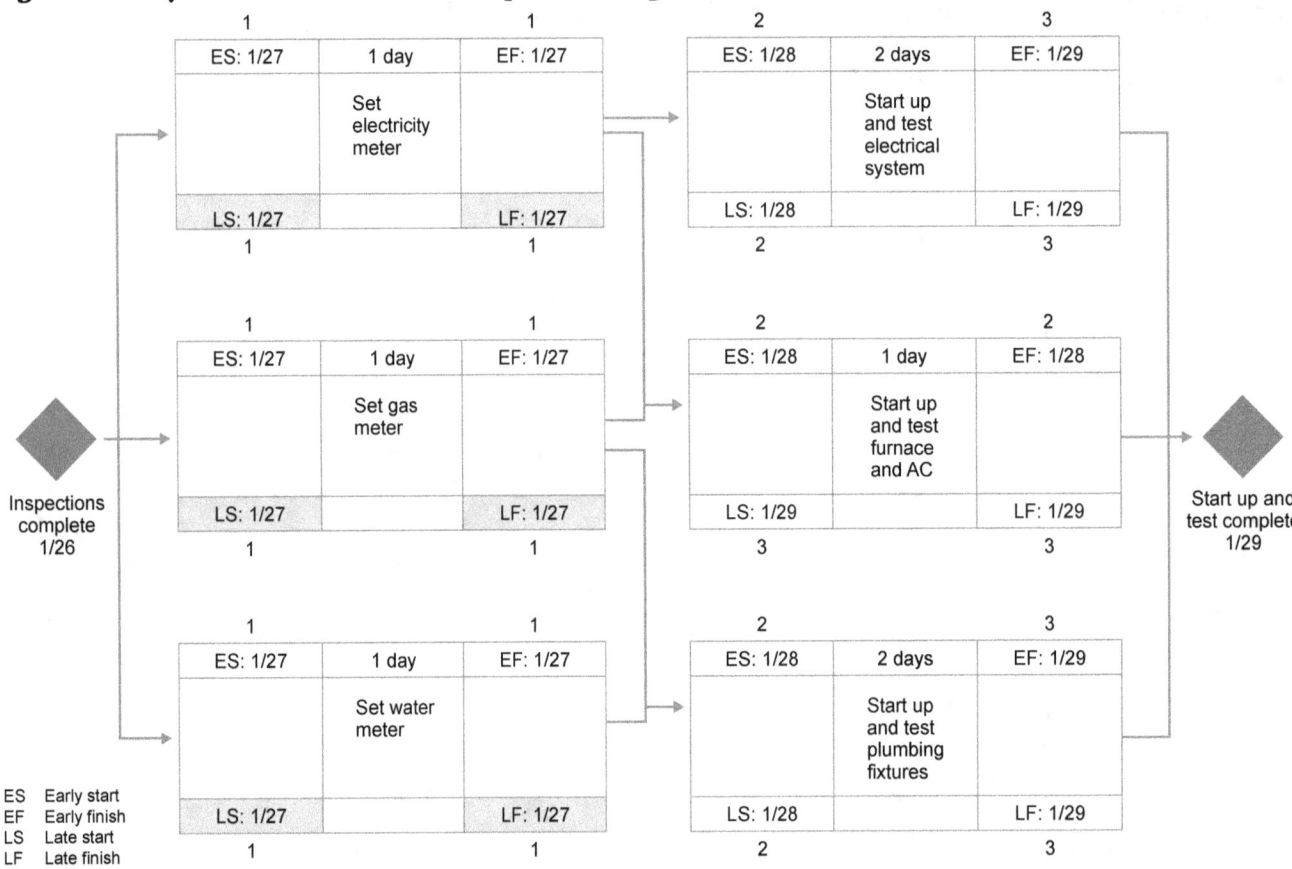

Source: GAO. | GAO-16-89G

TOTAL FLOAT

Once the early and late dates have been derived, the schedule can be assessed for flexibility. The difference between the time an activity may start or finish and the time it must start or finish in order for the project to be completed on time is known as total float (TF). Total float is calculated as the difference between an activity's early and late dates:

$$TF = LS - ES$$

or

$$TF = LF - EF$$

Total float is noted in figure 57 as "TF." Activities that have no total float are highlighted in red.

Figure 57: Total Float in a Start-Up and Testing Network

	1		1
ES: 1/27	1 day	EF: 1/27
TF 0 days	Set electricity meter	
LS: 1/27		LF: 1/27
	1		1

	2		3
ES: 1/28	2 days	EF: 1/29
TF 0 days	Start up and test electrical system	
LS: 1/28		LF: 1/29
	2		3

	1		1
ES: 1/27	1 day	EF: 1/27
TF 0 days	Set gas meter	
LS: 1/27		LF: 1/27
	1		1

	2		2
ES: 1/28	1 day	EF: 1/28
TF 1 day	Start up and test furnace and AC	
LS: 1/29		LF: 1/29
	3		3

	1		1
ES: 1/27	1 day	EF: 1/27
TF 0 days	Set water meter	
LS: 1/27		LF: 1/27
	1		1

	2		3
ES: 1/28	2 days	EF: 1/29
TF 0 days	Start up and test plumbing fixtures	
LS: 1/28		LF: 1/29
	2		3

Inspections complete 1/26

Start up and test complete 1/29

ES   Early start
EF   Early finish
LS   Late start
LF   Late finish
TF   Total float

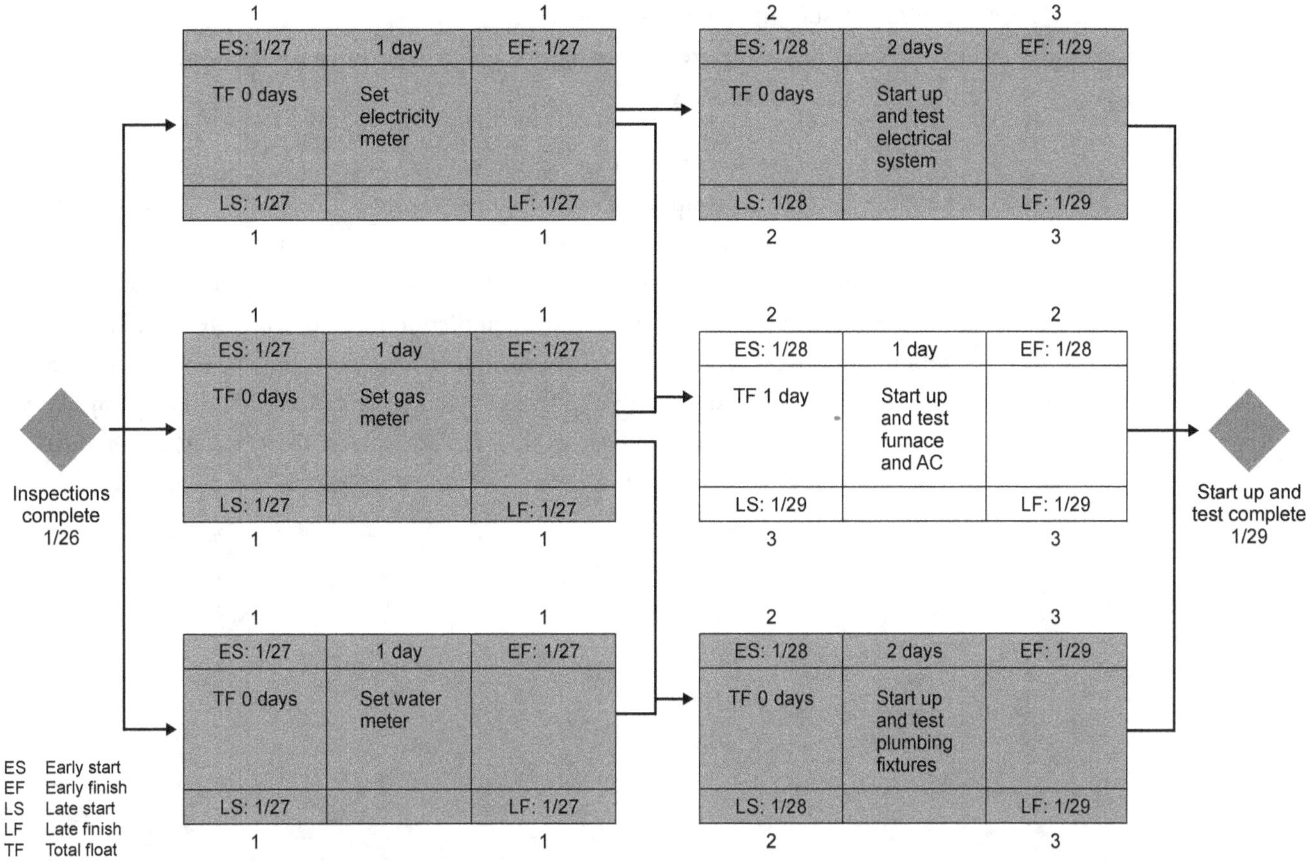

Source: GAO. | GAO-16-89G

All activities but one in figure 57 have 0 total float. A total float value of 0 indicates that an activity has no flexibility between the date when it may start and the date when it must start or between the dates when it may finish and must finish. Any delay in its start or finish dates transfers directly to the end milestone.

Three paths through the set-up and testing network have no total float and thus no flexibility:

1. "Inspections complete" → "set electricity meter" → "start up and test electrical system" → "start up and test complete";
2. "Inspections complete" → "set gas meter" → "start up and test plumbing fixtures" → "start up and test complete";
3. "Inspections complete" → "set water meter" → "start up and test plumbing fixtures" → "start up and test complete."

Any delay along the activities on these paths is transferred directly, day for day, to the finish date of the "start up and test complete" milestone, unless the delay is mitigated.

Two paths through the set-up and testing network have total float available and thus have flexibility:

1. "Inspections complete" → "set electricity meter" → "start up and test furnace and AC" →"start up and test complete";
2. "Inspections complete" → "set gas meter" → "start up and test furnace and AC" → "start up and test complete."

Practically speaking, only one activity in the network can slip a day. None of the meter-setting activities can slip because they are all on the critical path: a slip in any one of the three will cause either the electrical system or the plumbing fixtures test to slip a day, causing the finish milestone to slip a day. Only the furnace and AC test activity can safely slip 1 day without pushing the completion milestone out. To introduce more flexibility into the schedule, the general contractor may want to assign an extra electrician to the electrical system testing. An extra resource can reduce the duration by 1 day and therefore introduce an extra day of total float for that activity. Because total float is shared along a path, the test's predecessor, "set electricity meter," would also gain a day of total float.

APPENDIX V

COMMON NAMES FOR SCHEDULE DATE CONSTRAINTS AND THEIR EFFECTS

The names for specific date constraints differ in scheduling software. In this guide, we refer to date constraint names as defined by the extensible markup language (XML) schemas used by the DOD Performance Assessments and Root Cause Analyses (PARCA) office to standardize its collection of cost and schedule data. These data are collected in support of Integrated Program Management Report (IPMR) DI-MGMT-81861. The IPMR is used to collect data for measuring cost and schedule performance on DOD acquisition contracts.

Through a joint industry and government working group, DOD PARCA has developed a data exchange instruction (DEI) for the XML schema. The DOD DEI supplements the United Nations Center for Trade Facilitation and E-business (UN/CEFACT) (XML) schema *09B*. UN/CEFACT XML schemas enable entities involved in the execution of a project to exchange trade, schedule, and cost data throughout the life of the project with a standardized data content framework.

Table 10 presents the DOD/UN/CEFACT standard date constraint names along with alternative names given to the constraints in common scheduling software packages. Regardless of how one chooses to refer to a date constraint, it is far more important to recognize the effects of the constraint on the calculations within the schedule network.

Table 10: Common Names for Date Constraints and Their Primary Effects

DOD/UN/CEFACT standard	Alternative name	Primary effect
Finish no earlier than (FNET)	Early finish Finish on or after Finish not earlier than	Forward pass: if necessary, sets the early dates of the activity so that the early finish equals the constraint date
Finish no later than (FNLT)	Late finish Finish on or before Finish not later than	Backward pass: if necessary, sets the late dates of the activity so that the late finish equals the constraint date
Start no earlier than (SNET)	Early start Start on or after Start not earlier than	Forward pass: if necessary, sets the early dates of the activity so that the early start equals the constraint date
Start no later than (SNLT)	Late start Start on or before Start not later than	Backward pass: if necessary, sets the late dates of the activity so that the late start equals the constraint date
Must start on (MSON)	MSO Mandatory start	Always sets both early and late start dates equal to the constraint date

DOD/UN/CEFACT standard	Alternative name	Primary effect
Must finish on (MFON)	MFO Mandatory finish	Always sets both early and late finish dates equal to the constraint date
(Not used)	As soon as possible	The default for forward scheduling; sets the early start date as early as possible
(Not used)	As late as possible	The default for backward scheduling; sets the early finish date as late as possible

Source: GAO analysis of NDIA Joint Industry/Government UN/CEFACT XML Working Group and DOD information | GAO-16-89G.

As soon as possible (ASAP) and as late as possible (ALAP) date constraints are not included in the DOD/UN/CEFACT standard but can be considered standard date constraints. ASAP date constraints are simply the default situation for forward scheduling, and ALAP date constraints are the default situation for backward scheduling. ALAP date constraints can be used in forward scheduling to force the activity to begin as late as possible. Their use in this case is rare, however, because the ALAP constraint immediately eliminates all available float for the activity. Depending on the scheduling software, the ALAP constraint may eliminate all available total float of successor activities as well.

In addition, table 10 does not include software-specific date constraints. For example, some scheduling software lets schedulers set an activity's early and late finish dates equal while still permitting the activity to have a planned finish later than the constraint date. This allows the network to calculate negative total float on sequences of activities without actually constraining dates. These date constraint features differ by software, and users should understand the effects of particular constraints on a schedule network.

APPENDIX VI

STANDARD QUANTITATIVE MEASUREMENTS FOR ASSESSING SCHEDULE HEALTH

An assessment of schedule best practices encompasses both qualitative and quantitative information. Qualitative information is provided by program questions such as those detailed in appendix II. These questions are related to the general policies in place and procedures undertaken to create and maintain the schedule. The quantitative assessment involves a detailed analysis of the schedule data to determine the overall health of the network. While the questions addressed by the data analysis are also covered in appendix II, the quantitative assessment often involves filters and detailed data metric definitions. These filters and definitions are in table 11 for each best practice.

No "pass-or-fail" thresholds or tripwires are associated with the measures. Measures are evaluated in context with qualitative program information and any documented justification. Moreover, severity of the errors or anomalies takes precedence over quantity because any error can potentially affect the reliability of the entire schedule network.

Table 11: Standard Data Measures for Schedule Best Practices

Best practice	Measure	Note
1. Capturing all activities	Measures in Best Practice 1 provide basic information on the scope of the schedule, such as number and types of activities and level of detail	
	Total number of activities, including total summary, hammock, milestone, and detail activities	Summary activities may or may not be present in the scheduling software
	Total number of remaining activities, including total summary, hammock, milestone, and detail activities	A remaining activity is any activity that is not complete. "Remaining" may be defined as (1) an activity with an actual start or no actual start and no actual finish or (2) any activity that is not 100 percent complete. Issues may arise with either definition. For instance, an activity may be noted as 100 percent complete and not have an actual finish date, or it may have actual start and finish dates but be less than 100 percent complete. Summary activities may or may not be present in the scheduling software
	If applicable, number of activities marked as both a milestone and summary activity	An activity cannot be both a summary and a milestone
	Number of activities with no descriptive name	May or may not be valid activities
	Ratio of detail activities to milestones	Provides a rough indicator of the level of planning detail in the schedule. While there is no specific threshold, one or two detailed activities per milestone is probably a very low level of detail, while 10 is probably highly detailed

Best practice	Measure	Note
	Number of activities not mapped to program or contractor work breakdown structure	
	Number of contractor activities not mapped to a SOW paragraph or similar information	Depending on the nature of the effort, an activity may not be mapped to the statement of work
	Number of activities with duplicate names	Activity names should be unique and descriptive
2. Sequencing all activities	Best Practice 2 includes more advanced measurements to assess the reliability of the network logic. Thresholds for measures are not provided because, in theory, any missing or inappropriate logic may disrupt the entire network. The assessment of this best practice is related to the assessment of Best Practices 5, 6, and 7. If major deficiencies are identified in Best Practice 2, then a valid critical path, total float, and horizontal traceability are not possible. For minor deficiencies, an assessment of the schedule's critical path, total float, and response to tests of horizontal traceability are essential to understanding the implications of constraints and incorrect or missing logic. All activities in a schedule, regardless of detail or planning period, are subjected to this best practice	
	Number of remaining detail activities and milestones missing predecessor links	Does not include the start milestone; missing links to external activities (activities outside the scope of the current schedule file) may be excluded when a schedule is evaluated outside the IMS network
	Number of remaining detail activities and milestones missing successor links	Does not include the finish milestone; missing links to external activities (activities outside the scope of the current schedule file) may be excluded when a schedule is evaluated outside the IMS network
	Number of remaining detail activities and milestones missing both predecessor and successor links	
	Dangling activities: number of remaining detail activities and milestones with no predecessor on start date	Milestone activities may be excluded because their start and finish dates are the same; missing links to external activities (activities outside the scope of the current schedule file) may be excluded when a schedule is evaluated outside the IMS network; activities that have actually started may be excluded because their start dates have been determined
	Dangling activities: number of remaining detail activities and milestones with no successor off finish date	Milestone activities may be excluded because their start and finish dates are the same; missing links to external activities (activities outside the scope of the current schedule file) may be excluded when a schedule is evaluated outside the IMS network
	Percentage of logic links that are finish-to-start	The majority of relationships within a detailed schedule should be finish-to-start
	Number of remaining detail activities and milestones with start-to-finish links	Count either successor or predecessor links but do not count both. An S–F link is between two activities but represents only one link
	Number of remaining summary activities with logic links	May also be measured as "logic links to and from remaining summary activities," although this may be a different number
	Remaining detail activities and milestones with a great many predecessors	Assesses the schedule for path convergence. A relatively high number of predecessors may indicate a high-risk area. Note that not all predecessors are driving; only predecessors that have zero or low float have the ability to delay the successor when they are delayed
	Remaining detail activities and milestones with soft date constraints	

Best practice	Measure	Note
	Remaining detail activities and milestones with hard date constraints	
	Remaining detail activities and milestones with active SNET date constraints	If an activity's scheduled start date is the same as the SNET date, then the SNET constraint is more than likely preventing the activity from starting early. This is considered an active constraint. If an SNET constraint is earlier than the activity's start date, then the activity is not affected by the constraint date
	Remaining detail activities and milestones with active FNET date constraints	If an activity's scheduled finish date is the same as the FNET date, then the FNET constraint is more than likely preventing the activity from finishing early. This is considered an active constraint. If an FNET constraint is earlier than the activity's finish date, then the activity is not affected by the constraint date
	Remaining detail activities and milestones with lags	Count either successor or predecessor lags but not both. A lag is between two activities but represents only one lag. This number is different from the number of lags
	Number of lags on remaining detail activities and milestones	Count either successor or predecessor lags but not both. A lag is between two activities but represents only one lag. This number is different from the number of activities with lags
	Remaining detail activities and milestones with leads	Count either successor or predecessor leads but not both. A lead is between two activities but represents only one lead. This number is different from the number of leads
	Number of leads on remaining detail activities and milestones	Count either successor or predecessor leads but not both. A lead is between two activities but represents only one lead. This number is different from the number of activities with leads
	Remaining detail activities and milestones with an F–S predecessor lead greater than remaining duration	
3. Assigning resources to all activities	Best Practice 3 is more programmatic than quantitative, although measures and trends may be investigated for fully resource loaded schedules. If possible, resource assignments over time may be evaluated to identify potential unrealistic peaks. In general, the measures assess the number of activities within the detail planning period that are assigned resources and the reasonableness of work hours. Overallocated resources and unrealistic resource units should be a cause for concern. Care should be taken to assess only the appropriate detailed activities	
	Total number of resources	
	Overallocated resources	
	Maximum units available per resource	Individuals should be available between 0 and 100 percent of full time, and resource groups should have a realistic number of individuals available to perform the work.
	Summary activities and milestones with assignments	Summary activity durations depend on the activities contained within them. Milestones should never be assigned resources because they have no duration
	Remaining detail activities with assignments	Exclude nonapplicable activities such as planning packages and reference activities
	Remaining detail activities without assignments	

Best practice	Measure	Note
4. Establishing the durations of all activities	Measures for Best Practice 4 are generally straightforward, providing an overall assessment of the detail available to management, as well as the appropriateness of the schedule calendars. Care should be taken to assess only the appropriate detailed activities	
	Remaining detail activities with dissimilar time units	All durations should be in the same unit, preferably days
	Remaining detail activities or milestones starting or finishing on a weekend or holiday	May be legitimate but may stem from incorrect calendar assignments or specifications. Milestones on weekends or holidays should be questioned
	Holidays and other exceptions by task calendar	
	Remaining detail activities with durations less than the reporting period	Exclude nonapplicable activities such as planning packages and LOE and reference activities. The analyst should take into account baseline durations if available. That is, if the baseline duration is 35 days but the actual plus remaining duration is 60, the original baseline meets the intent of the best practice
	Remaining detail activities with durations greater than the reporting period	
	Average duration of remaining detail activities	
	Median duration of remaining detail activities	
5. Verifying that the schedule can be traced horizontally and vertically	Best Practice 5 has no standard measurements. Vertical traceability is assessed by determining whether lower-level activities fall within the same time as higher-level activities and whether detailed schedule dates fall within the same time as summary schedule dates. An essential check of vertical traceability is determining whether forecasted milestone dates in detailed schedules match those quoted in management documents. Horizontal traceability depends on Best Practice 2, although not entirely as noted in that best practice. It is assessed by increasing activities' durations by improbable amounts (500 or 1,000 days) and by observing how the schedule reacts. In the absence of constraints and assuming logic has been properly identified, key milestones should move and the critical path should change	
	Assessment of how critical and noncritical planned dates dynamically react to dramatic increases in predecessor activity durations	Horizontal traceability implies that the network responds dynamically to delayed activities. Severely delayed activities should become critical and previously critical paths should become noncritical. Delays of this magnitude should cause the finish date to slip relative to the activity delay and reasonable available float
6. Confirming that the critical path is valid	Best Practice 6 has several standard measurements for assessing the validity of a critical path. Beginning at the program finish milestone, the sequence of driving activities is traced back to the status date. This sequence of activities should be straightforward, continuous, and the same as the critical path—defined by zero total float—in the absence of date constraints. Critical paths to interim key milestones may also be assessed as applicable	
	Assessment of the driving paths to key milestones and comparison of those paths to activities marked as critical in the schedule	Ideally the longest path and critical path are the same to the key milestone. The path should be continuous from the status date to the key milestone
	Number of critical activities	In general, if the ratio of critical path activities to the total remaining activity count is nearly 100 percent, then the schedule may be overly serial and resource limited
	Number of critical LOE activities	A critical path cannot include LOE activities because they do not represent discrete effort
	Number of lags and leads on the critical path	Lags cannot represent work and cannot be assigned resources
	Number of critical activities with hard date constraints	Using hard constraints to fix activity dates at certain points in time immediately convolutes critical path calculations and defeats the purpose of CPM scheduling

Best practice	Measure	Note
	Number of in-progress critical activities	Given that the critical path is a continuous sequence of activities from the status date, at least one in-progress activity is critical
7. Ensuring reasonable total float	Best Practice 7 includes basic measurements of total float to assess overall program flexibility as reported by the schedule. It is closely related to assessments of Best Practices 2, 5, and 6, because a properly sequenced network produces reasonable estimates of float and a valid critical path. Reasonableness is assessed in combination with program length and activity type. In addition, because one logic error can cause an entire sequence of activities to report unreasonable amounts of float, the breadth of deficiencies reported in Best Practice 2 should be taken into account here. Negative float should always be questioned	
	Remaining detail activities and milestones with dissimilar total float time units	All float should be in the same units, preferably days
	Remaining detail activities and milestones with relatively high total float	High float is relative to the scope, length, and complexity of the schedule. Float should be reasonable and should realistically reflect the flexibility of the schedule
	Remaining detail activities with negative total float	Negative total float indicates that the activity's constraint date is earlier than its calculated late finish. Negative float may occur when activities are performed out of sequence
	Average total float value of remaining detail activities and milestones	
	Median total float value of remaining detail activities and milestones	
8. Conducting a schedule risk analysis	Many quantitative measurements are related to Best Practice 8, and a proper schedule risk analysis typically deserves a much more complex quantitative assessment than that given here. GAO's assessment of Best Practice 8 is more programmatic, and these questions are provided in appendix II. The measures for Best Practice 8 are limited to determining the existence of risk data within the schedule risk file	
	Fields within the schedule used for SRA	Fields that store optimistic, most likely, and pessimistic durations
	Correlation measures within the schedule	
	Contingency activities	
9. Updating the schedule using actual progress and logic	Best Practice 9 is assessed by determining the validity of the dates reported in the schedule. The assessment of this best practice depends on the status date reported in the schedule. It also depends on the scheduling software used: some software packages allow date anomalies that other software packages prevent	
	Number of in-progress activities	An activity is in progress when it has started but is not yet complete
	Number of remaining detail activities and milestones that have a forecasted start date in the past but no actual start date	Forecasted start dates should not occur in the past—i.e., any time preceding the status date
	Number of remaining detail activities and milestones that have a forecasted finish date in the past but no actual finish date	Forecasted finish dates should not occur in the past—i.e., any time preceding the status date
	Number of remaining detail activities and milestones that have an actual start date in the future	Actual start dates should not occur in the future—i.e., any time following the status date
	Number of remaining detail activities and milestones that have an actual finish date in the future	Actual finish dates should not occur in the future—i.e., any time following the status date
	Number of detail activities performed out of sequence	

Best practice	Measure	Note
10. Maintaining a baseline schedule	Many data measures can be used to assess Best Practice 10; some are provided here. All baseline measures ultimately depend on the existence of a controlled baseline and a properly statused current schedule. Baseline measures are typically calculated by reporting period: for example, number of activities forecast to start early over the next 60 days or activities that have actually finished late over the past 6 months. They may also be useful when applied to specific products within the WBS, resource groups, or criticality: for example, the number of late activities during product integration, the average start variance of activities executed in one production plant, or the baseline execution index of activities with less than 10 days of total float	
	Number of detail activities and milestones with baseline dates	Counts should accord with rolling wave planning periods
	Number of detail activities and milestones without baseline dates	
	Number of remaining detail activities and milestones that are forecast to start or finish before their baseline dates	Represents activities and milestones forecast to begin or end early
	Number of remaining detail activities and milestones that are forecast to start or finish on their baseline dates	Represents activities and milestones forecast to begin or end on time
	Number of remaining detail activities and milestones that are forecast to start or finish after their baseline dates	Represents activities and milestones forecast to begin or end late
	Number of remaining detail activities that actually started before their baseline start date	Represents activities and milestones that actually started early
	Number of remaining detail activities that actually started on their baseline start date	Represents activities and milestones that actually started on time
	Number of remaining detail activities that actually started after their baseline start date	Represents activities and milestones that actually started late
	Number of detail activities and milestones that actually finished before their baseline finish date	Represents activities and milestones that actually finished early
	Number of detail activities and milestones that actually finished on their baseline finish date	Represents activities and milestones that actually finished on time
	Number of detail activities and milestones that actually finished after their baseline finish date	Represents activities and milestones that have actually finished late
	Average and median start variance	Start variance may be the difference between actual start and baseline start or forecast start and baseline start
	Average and median finish variance	Finish variance may be the difference between actual finish and baseline finish or forecast finish and baseline finish
	Baseline execution index	The ratio of actual completed detail activities to detail activities that were planned to finish. A BEI of 1 indicates that the project is performing according to plan. A BEI less than 1 indicates that, in general, fewer activities are being completed than planned; a BEI greater than 1 indicates that, in general, more activities are being completed than planned

Source: GAO | GAO-16-89G.

APPENDIX VII

COMPARISON OF GAO's SCHEDULE ASSESSMENT GUIDE TO KEY INDUSTRY AND AGENCY SCHEDULE GUIDANCE

We compared information described in the GAO *Schedule Assessment Guide* to several major sources of schedule guidance. This appendix summarizes the extent to which we identified comparable information in those sources as well as substantive gaps in best practices.

In general, we found that best practices described in the *Schedule Assessment Guide* agree with best practices in the following sources of guidance. One recurring difference is the requirement to assign resources to activities in the schedule. All but one of the schedule guidance documents reviewed below do not require assigning resources to activities. Nevertheless, the documents describe many benefits associated with performing the practice. Only DHS's *Scheduling Handbook* states that an IMS should be resource loaded. In addition, any schedule adhering solely to DoD's IMS Data Item Description (DID) 81650 or Integrated Program Management Report (IPMR) DID 81861 will not meet all best practices described in the Schedule Guide. Neither DID requires government activities be integrated with contractor activities or that resources be assigned to activities.

The guidance documents below contain information beyond what is covered in the GAO *Schedule Assessment Guide*. For example, they include guidance on how master schedules are used throughout an agency's acquisition process; additional information on tools and metrics; and advanced topics such as business rhythm and joint confidence levels. We were able to enhance and improve many best practices described in the GAO *Schedule Assessment Guide* by conducting comparison analyses with these reputable guides.

DOD DEFENSE CONTRACT MANAGEMENT AGENCY

In its role as the DOD executive agent for earned value management systems (EVMS), the Defense Contract Management Agency's (DCMA) mission includes conducting contractor surveillance on EVMS.[46] The outcome of DCMA surveillance ensures that reported contract performance data accurately reflect the status of programs.

---

[46]We referenced the following instructions and guidelines: *Earned Value Management System Compliance Reviews Instruction* (DCMA-INST 208); *Overview: 14 Point Assessment* (EVC-104_Rev 1); IMS assessment guides (EVC-101_Rev11, EVC-102_Rev8, and EVC-103_Rev7); *Schedule Margin Position Paper* (EVC-106_Rev2); and *Finding the Critical Path* (EVC-100_Rev1).

In assessing contractor schedule reliability, DCMA ensures that the following ANSI/EIA-748 EVMS guidelines are followed:

- Schedule the authorized work to describe the sequence of work and identify significant task interdependencies required to meet the requirements of the program.
- Identify physical products, milestones, technical performance goals, or other indicators that will be used to measure progress.

In its assessment of the quality of a schedule, DCMA uses a 14-Point Assessment (14PA), a collection of measures intended to assess the technical structure of the schedule as well as the contractor's ability to plan and execute work. The measures are:

1. Logic
2. Leads
3. Lags
4. Relationship types
5. Hard constraints
6. High float
7. Negative float
8. High duration
9. Invalid dates
10. Resources
11. Missed tasks
12. Critical path test
13. Critical path length index
14. Baseline execution index.

Several include thresholds; for example, no more than 5 percent of remaining tasks should be missing predecessor or successor logic. However, DCMA's 14PA thresholds are not compliance triggers. Rather, they are used as a starting point toward an objective analysis of the schedule.

A GAO and a DCMA schedule assessment have inherently different purposes. Notably, DCMA's review focuses on contractor adherence to ANSI standards and contractual data deliverables. This difference affects to some extent the applicability of best practices discussed in the *Schedule Assessment Guide*. For example, when assessing whether a schedule includes all effort, DCMA ensures that only scope that is on contract is included in the contractor's schedule. In addition, DCMA ensures that resources are properly loaded in a schedule only if a contract requires resource-loaded schedules. Finally, schedule risk analyses (SRA) are not included as part of the EVMS guidelines. Therefore, DCMA does not assess whether an SRA has been performed unless one is required by contract.

Otherwise, we found few substantive differences between best practices detailed in the *Schedule Assessment Guide* and the DCMA 14PA and other DCMA documentation.

Salient differences are that, first, DCMA assessments allow unlimited use of soft date constraints, while the *Schedule Assessment Guide* recommends minimizing and justifying their use. Second, DCMA procedures do not describe guidelines for ensuring that SRAs are conducted properly even if one is required by contract. Finally, DCMA guidelines do not recommend the use of a schedule basis document or a schedule narrative.

Given DCMA's focus on ensuring the validity of schedules in an earned value management environment, in some cases DCMA procedures and measures go beyond best practices described in the *Schedule Assessment Guide*. These DCMA measures focus on ensuring that control account budgets are valid, cost and schedule estimates are fully integrated, earned value techniques are validated, and the like.

DOD NAVAL AIR SYSTEMS COMMAND COST DEPARTMENT, INTEGRATED PROJECT MANAGEMENT DIVISION

DOD's Naval Air Systems Command Cost Department (NAVAIR 4.2) conducts cost estimates and analysis throughout the life cycle of naval aviation programs and their affiliates. NAVAIR 4.2 is organized into three divisions. NAVAIR 4.2.3, Integrated Project Management, is responsible for, among other things, IMS development and maintenance, schedule risk assessments, integrated baseline reviews, and earned value and schedule analysis. The Integrated Project Management Toolkit and associated Schedule Metrics Guide (referred to here as the Toolkit) are used to provide NAVAIR 4.2.3 analysts guidance for validating contractor master schedules and for preparing monthly analyses for government program management.[47]

To analyze a schedule, NAVAIR uses an 11-point assessment described in the Toolkit. Its assessment steps are

1. Completeness of the IMS
2. Critical target dates established and used for planning
3. Sequence of work
4. Schedule architecture and integration
5. Proper use of constraints, leads, and lags
6. Necessary and consistent detail for tasks
7. Resource adequacy and availability
8. Critical path calculation and reasonableness
9. Reasonableness of slack
10. IMS status and forecasting abilities
11. Evaluation of the risk in the IMS.

---

[47]We referenced the following instructions and guidelines: *Naval Air Systems Command Integrated Project Management Toolkit;* Naval Air Systems Command *Schedule Metrics Guide;* Naval Air Systems Command *Schedule Risk Assessment Toolkit.*

Best practices described in the Toolkit differ in several ways from best practices in the GAO *Schedule Assessment Guide.* The Toolkit does not verify whether government activities are fully integrated with contractor activities or whether activities are resource loaded. The Toolkit does not address these best practices because the NAVAIR 11-point assessment is an interpretation of ANSI/EIA-748 guidelines, the intent of the IMS DID 81650, and the IPMR DID 81861. As NAVAIR notes, the assessment helps verify that the contractor's scheduling techniques meet the minimum requirements of network scheduling. Focusing on a contractor's IMS follows DOD's practice of defining an IMS in terms of contracted work.

However, NAVAIR does promote the integration of contractor and government activities through links between the contractor IMS and the integrated government schedule (IGS). The IGS is a government-controlled schedule that contains activities the government needs to perform in support of a program. The IGS also includes givers and receivers that are linked to the contractor IMS. While an IGS is not required, a program management office may request one; in this case, NAVAIR 4.2 analysts will assist in creating and maintaining the IGS and its links to the contractor IMS.

Finally, because neither DID 81650 nor DID 81861 requires that resources be assigned to activities in an IMS, the Toolkit has no measurements or process to verify the schedule's activities are assigned resources. However, NAVAIR currently has an initiative to develop a process for resource loading the IGS.

We found no other substantive differences between best practices detailed in the *Schedule Assessment Guide* and the Toolkit and other NAVAIR documentation. The Toolkit describes 76 schedule measures, some of which are not covered by the *Schedule Assessment Guide.* Toolkit measures are accompanied by extensive documentation describing their definitions, formulas or filters, purpose, examples, and cautions to be aware of while interpreting results. Most of the measures not covered by the *Schedule Assessment Guide* are detailed comparisons and trend analyses of the current schedule against the baseline schedule. Similar to GAO's criteria, measurements in the Toolkit do not rely on tripwires or thresholds. Rather, a schedule's health is determined by assessing the impact of schedule anomalies on the network.

DEPARTMENT OF HOMELAND SECURITY'S *SCHEDULING HANDBOOK*

DHS's Program Management Center of Excellence developed the *Scheduling Handbook* to describe (1) scheduling best practices DHS and other agencies use; (2) scheduling techniques and tools for creating reliable schedules; and (3) how the creation and maintenance of schedules coincide with DHS's acquisition framework, including acquisition decision events.[48] The handbook contains a scorecard for evaluators to ensure that the

---

[48] *Scheduling Handbook,* August 29, 2014.

IMS meets the four characteristics of a reliable schedule: comprehensive, well-constructed, credible, and controlled. DHS's Program Accountability and Risk Management (PARM) office reviews the scorecard before an acquisition decision event. In addition to reviewing the IMS, PARM also reviews the schedule narrative, baseline schedule document, the WBS, and other necessary documentation.

Best practices defined in the handbook are designed to align with the ten best practices described in GAO's *Schedule Assessment Guide*. We found no substantive differences between them.

NATIONAL AERONAUTICS AND SPACE ADMINISTRATION'S *SCHEDULE MANAGEMENT HANDBOOK*

NASA's *Schedule Management Handbook* is guidance for NASA headquarters, centers, laboratories, partners, and contractors.[49] The purpose of the handbook is to provide guidance on meeting NASA's schedule requirements and to describe best practices, concepts, and techniques associated with schedule management. The handbook also details the importance of scheduling in NASA's program life-cycle management process.

We found few differences between best practices detailed in the *Schedule Management Handbook* and GAO's *Schedule Assessment Guide*. The *Schedule Management Handbook* focuses on ensuring that all approved and authorized scope is included in the schedule but also states that master schedules should model the total integrated plans for the entire program.

NASA guidance does not require resource loaded schedules, but the handbook does emphasize the importance of loading and assigning resources. The handbook states that resources should be assigned within the schedule itself to ensure proper cost and schedule integration. In addition, NASA guidance stresses that resource loading and leveling is recommended to ensure that the plan is complete and credible; otherwise, significant risk is assumed if a schedule is baselined without first being resource loaded and leveled. The handbook also states that resources can be managed externally to the schedule using a spreadsheet, but it does not describe a process for integrating information this way or for resolving resource conflicts.

NASA's *Cost Estimating Handbook*[50] describes NASA's joint confidence level (JCL), a concept not covered by GAO's *Schedule Assessment Guide*. The JCL is a quantitative probability analysis that requires a project to combine its cost, schedule, and risks into a complete quantitative picture to help assess whether a project will be successfully completed within cost and on schedule. NASA introduced the analysis in 2009, and it is among the agency's initiatives to reduce acquisition management risk. NASA's proce-

---

[49]*Schedule Management Handbook,* NASA/SP-2010-3403, March 2011.

[50]*Cost Estimating Handbook* Version 4.0, February 2015.

dural requirements state that mission directorates should plan and budget programs and projects based on a 70 percent JCL or at a different level as approved by the Decision Authority of the Agency Program Management Council, and any JCL approved at less than 70 percent must be justified and documented.

NATIONAL DEFENSE INDUSTRIAL ASSOCIATION *PLANNING AND SCHEDULING EXCELLENCE GUIDE*

The National Defense Industrial Association, made up of government, industry, and academia members, is a forum for the exchange of national security information between industry and government. The association published the *Planning and Scheduling Excellence Guide* (PASEG) in June 2012.[51] The intent of the PASEG is to provide program management teams guidance on creating, maintaining, and analyzing an integrated master schedule using a disciplined schedule and planning process. It includes sections on schedule architecture, modeling techniques, resource integration, schedule maintenance, schedule analysis, and guidance on scheduling for specific contract phases and production environments.

The PASEG is described as a reference whose approaches and techniques should be implemented only if program management deems them achievable and necessary. The PASEG states that practices described in the guide apply to any industry but that its primary audience is large defense and intelligence program management teams. The PASEG is intended as supplemental guidance to, and states that it is subordinate to, IMS DID 81650, the IPMR DID 81861, and a contractor's management procedures.

Hence, comparisons between the GAO *Schedule Assessment Guide* and the PASEG are similar to comparisons with other schedule guidance documents that are compliant with, supplemental to, or subordinate to DIDs 81650 and 81861. GAO's *Schedule Assessment Guide* and the PASEG have a few minor differences but in general the two documents describe similar best practices and procedures for creating and maintaining reliable schedules. The primary difference between the guides—owing to subordination to DIDs 81650 and 81861—is that an IMS is not required to be resource loaded and need consist only of contracted work.

At the same time, however, and in agreement with the Schedule Guide, the PASEG describes integrating all work and assigning resources as good practices. First, the PASEG describes concepts that agree with best practices in the GAO *Schedule Assessment Guide* for capturing all work and integrating government and contractor activities. For example, the PASEG notes that the IMS should reflect all work, and capture all activities, required to complete the program. In addition, it states that the customer schedule can be integrated with the IMS to ensure a comprehensive view of the remaining work for the program.

---

[51] *Planning and Scheduling Excellence Guide* Release 2.0, June 22, 2012.

Second, the PASEG describes concepts that, for the most part, agree with best practices in the GAO *Schedule Assessment Guide* for assigning resources to all activities. The document describes three approaches to integrating resources with schedule activities: assigning resources to activities in the schedule; aligning resources from an external cost system to activities in the schedule; and using text fields in the schedule software to note resource information. In agreement with the Schedule Guide, the PASEG states that the first method—maintaining resources in the schedule—is the recommended approach because it provides information on resource requirements and serves as the focal point for cost and schedule integration. PASEG guidance and the Schedule Guide also agree on the second method, noting that integration with external cost systems may be necessary in some cases. However, the third method—integrating resources using text fields in the schedule—is not discussed in the Schedule Guide and GAO does not consider it a best practice. Notably, the PASEG states that, regardless of the method used to integrate resources into the schedule, every activity should have assigned resources, and program management should derive resource requirements from the IMS.

The PASEG also includes eight principles for creating and maintaining a sound schedule. It states that the schedule team's meeting these principles ensures that it has created and maintains a robust IMS using a rigorous and disciplined process. The principles are collectively called the generally accepted schedule principles, or GASP. We found that all but one GAO best practices map to at least one principle: Best Practice 8, Schedule Risk Analysis, has no corresponding generally accepted schedule principle (table 12).

Table 12. Generally Accepted Scheduling Principles and GAO Best Practices Compared

Generally accepted scheduling principle	Description	Corresponding GAO best practice
Complete	The schedule captures the entire discrete, authorized project effort from start through completion.	1. Capturing all activities
Traceable	The schedule logic is horizontally and vertically integrated with cross-references to key documents and tools. Schedules are coded to relate tasks and milestones to documents and responsible organizations.	1. Capturing all activities 2. Sequencing all activities 5. Verifying that the schedule can be traced horizontally and vertically
Transparent	The schedule provides visibility to ensure that it is complete, is traceable, has documented assumptions, and provides full disclosure of program status and forecast.	9. Updating the schedule using actual progress and logic 10. Maintaining a baseline schedule
Statused	The schedule shows accurate progress through the status date.	9. Updating the schedule using actual progress and logic
Predictive	The schedule provides meaningful critical paths and accurate forecasts for remaining work through program completion.	4. Establishing the durations of all activities 6. Confirming that the critical path is valid 7. Ensuring reasonable total float
Usable	The schedule is an indispensable tool for timely and effective management decisions and actions.	1. Capturing all activities 10. Maintaining a baseline schedule

Generally accepted scheduling principle	Description	Corresponding GAO best practice
Resourced	The schedule aligns with actual and projected resource availability.	3. Assigning resources to all activities
Controlled	The schedule is built, baselined, and maintained using a stable, repeatable, and documented process.	10. Maintaining a baseline schedule

Source: GAO analysis of NDIA information | GAO-16-89G.

The PASEG also describes the benefits of performing schedule assessments and provides a list of the most common measures. Similar to GAO's best practices, measures in the PASEG do not rely on tripwires or thresholds. Rather, a schedule's health is determined by assessing the effect of schedule anomalies on the network.

APPENDIX VIII

RECOMMENDED ELEMENTS OF A DATA COLLECTION INSTRUMENT

1. The baseline IMS and the latest updated IMS, including all applicable embedded project schedules. Their format should be that of a software schedule file. PDF files and presentation slides are not valid schedule file formats. The name and version of the software used to create and maintain the schedule are provided.

2. A schedule dictionary or similar documentation that defines custom fields, especially those that contain information on level-of-effort activities; contractor versus government effort; and statement of work, statement of objective, work package, integrated master plan, or control account mappings.

3. Integrated master plan (IMP), if applicable.

4. Work breakdown structure (WBS) and dictionary.

5. A statement of work (SOW), if applicable.

6. A cross-walk between the WBS, contractor WBS, SOW and the schedule activities, as applicable.

7. An identification of the main deliverables, including a designation of the paths that the project considers critical and near-critical.

8. Schedule basis documentation.

9. Schedule narrative documentation.

10. A basis of estimate or other documentation for estimating activity durations and assigned resources.

11. Program management review briefings or similar documentation discussing schedule status and establishing traceability of reported dates to detail schedules.

12. Relevant scheduling guidance, such as contract line item numbers, data item descriptions, and agency directives that govern the creation, maintenance, structure, and status of the schedule.

13. Schedule risk analysis documentation, including the analytical approach, assumptions, and results.

14. A risk management plan and a copy of the current risk register.

15. A description of the schedule change control process.

APPENDIX IX

CASE STUDY BACKGROUNDS

The material in the guide's 19 case studies was drawn from the 11 GAO reports described in this appendix. Table 13 shows the relationship between reports, case studies, and the chapters in which they are cited. The table is arranged by the order in which we issued the reports, earliest first. Following the table, paragraphs describe the reports, ordered by the numbers of the case studies in the guide.

Table 13: Case Studies Drawn from GAO Reports Illustrating This Guide

Case study	GAO report	Best practices cited
1, 5, 6, 16	GAO-11-53: DOD Business Transformation	1, 2, 9
2, 10	GAO-14-152: DOD Business Systems Modernization	1, 5
3, 8	GAO-14-368: Arizona Border Surveillance Technology Plan	1, 3
4, 15	GAO-10-189: VA Construction	1, 8
7	GAO-10-378: Nuclear Nonproliferation	3
9	GAO-10-43: Transportation Worker Identification Credential	4
11, 13	GAO-12-223: FAA Acquisitions	6, 7
12	GAO-12-66: Immigration Benefits	6
14	GAO-11-743: Coast Guard	8
17	GAO-11-740: Aviation Security	9
18	GAO-13-676: Polar-orbiting Environmental Satellites	10
19	GAO-10-59: 2010 Census	10

Source: GAO | GAO-16-89G.

CASE STUDIES 1, 5, 6, AND 16: *FROM DOD BUSINESS TRANSFORMATION*, GAO-11-53, OCTOBER 7, 2010

The Department of Defense (DOD) invests billions of dollars annually to modernize its business systems, and the DOD business systems modernization program has been on our high-risk list since 1995. In its modernization, DOD is implementing nine enterprise resource planning (ERP) efforts that perform business-related tasks such as general ledger accounting and supply chain management. These efforts are essential to trans-forming DOD's business operations.

We were asked to determine whether selected ERPs followed schedule and cost best practices. We found that none of the four programs we assessed had developed a fully integrated master schedule as an effective management tool. Such a tool is crucial to es-timating the overall schedule and cost of a program. Without it, DOD is unable to say, with any degree of confidence, whether the estimated completion dates are realistic.

In particular, we found that selected project schedules for the Air Force's Expeditionary Combat Support System (ECSS) did not meet best practices. We recommended that the Secretary of the Air Force ensure that the Chief Management Officer of the Air Force direct the program management office for ECSS to develop an IMS that, among other things, sequenced all activities, assigned resources to all activities, and established a valid critical path. In November 2012, DOD cancelled the ECSS program, citing the lack of an IMS as a major cause of its failure.

CASE STUDIES 2 AND 10: FROM *DOD BUSINESS SYSTEMS MODERNIZATION*, GAO-14-152, FEBRUARY 7, 2014

The Air Force's Defense Enterprise Accounting and Management System (DEAMS) was initiated in August 2003 and is intended to provide the Air Force with the entire spectrum of financial management capabilities, including collections, commitments and obligations, cost accounting, general ledger, funds control, receipt and acceptance, accounts payable and disbursement, billing, and financial reporting for the general fund. DOD has stated that the development and implementation of DEAMS is critical to its goal of producing auditable financial statements by September 2017, as called for by the National Defense Authorization Act for Fiscal Year 2010.

To support the Congress's continuing oversight of DOD's progress in implementing its ERP systems, we reviewed the schedule and cost estimates for selected DOD ERP systems. The objective of this review was to determine the extent to which the current schedule and cost estimates for DEAMS were prepared in accordance with GAO's Schedule and Cost Guides. We reviewed the most current schedule and cost estimates that supported DOD's February 2012 Milestone B decision, which determined that investment in DEAMS was justified.

We found that the schedule for the DEAMS program did not meet best practices. The cost estimate did meet best practices, but the issues associated with the schedule could negatively affect the cost estimate. Specifically, the DEAMS schedule supporting the February 2012 Milestone B decision partially or minimally met the four characteris-tics for developing a high-quality and reliable schedule—it was not comprehensive, well-constructed, credible, or controlled.

In contrast, the DEAMS cost estimate fully or substantially met the four characteristics of a high-quality and reliable cost estimate—it was comprehensive, well-documented, accurate, and credible. However, because the cost estimate is based on the schedule, the unreliability of the schedule could affect the cost estimate. For example, if there are schedule slippages, the costs for the program could be greater than currently estimated.

CASE STUDIES 3 AND 8: FROM *ARIZONA BORDER SURVEILLANCE TECHNOLOGY PLAN*, GAO-14-368, MARCH 3, 2014

In recent years, nearly half of all annual apprehensions of illegal U.S. entrants along the southwest border have occurred along the Arizona border. Under the Secure Border Initiative Network (SBI*net*), the Department of Homeland Security's (DHS) Custom and Border Protection (CBP) deployed surveillance systems along 53 of the 387 miles of the Arizona border with Mexico. After DHS canceled further SBI*net* procurements, CBP developed the Arizona Border Surveillance Technology Plan, which includes a mix of radars, sensors, and cameras to help provide security for the remainder of Arizona's border. GAO was asked to review the status of DHS's efforts to implement the Plan. Our report addressed the extent to which CBP developed schedules and life-cycle cost estimates for the Plan in accordance with best practices; followed aspects of DHS's acquisition management guidance in managing the Plan's programs; and identified mission benefits and developed performance measures for deploying surveillance technologies under the Plan.

We obtained program schedules as of March 2013 that were current at the time of our review for the three highest-cost programs—Integrated Fixed Towers, Remote Video Surveillance System, and Mobile Surveillance Capability—and we compared the schedules with best practices for developing schedules outlined in an exposure draft of GAO's Schedule Assessment Guide. We also interviewed cognizant officials in CBP's Office of Technology Innovation and Acquisition (OTIA) and program offices. By assessing the schedules against best practices, we identified CBP's schedule challenges in testing, procuring, deploying, and operating technologies under the Plan. We interviewed CBP officials to determine reasons for the schedule challenges and steps that CBP had taken or was taking to address them.

CASE STUDIES 4 AND 15: FROM *VA CONSTRUCTION*, GAO-10-189, DECEMBER 14, 2009

The Department of Veterans Affairs (VA) operates one of the largest health care systems in the nation. As of August 2009, VA's Veterans Health Administration (VHA) had 32 major ongoing construction projects, with an estimated total cost of about $6.1 billion and average cost per project of about $191 million. Some of these projects were initiated as part of VA's Capital Asset Realignment for Enhanced Services (CARES), which was a comprehensive assessment of VHA's capital asset requirements. In response to a congressional request, we (1) described how costs and schedules of current VHA major construction projects had changed, (2) determined the reasons for changes in costs and schedules, and (3) described the actions VA had taken to address cost increases and schedule delays.

We reviewed construction documents, visited three construction sites, and interviewed VA and general contractor officials. We found that while about half of the 32 major on-going construction projects were within their budgets, since submitted to the Congress, 11 had schedule delays, 18 had been subjected to cost increases, and 5 showed a cost increase of over 100 percent. For example, the cost of a new medical center in Las Vegas rose from an initial estimate of $286 million to over $600 million, an increase of about 110 percent. In addition, 13 projects had been subjected to cost increases of between 1 and 100 percent and 11 to schedule delays, 4 of which were longer than 24 months. Several reasons for project cost increases and schedule delays included VA's preparing initial cost estimates that were not thorough, its making significant changes to project scope after initial estimates were submitted, and unforeseen events such as an increase in the cost of construction materials.

CASE STUDY 7: FROM *NUCLEAR PROLIFERATION*, GAO-10-378, MARCH 26, 2010

The end of the Cold War left the United States with a surplus of weapons-grade plutonium that posed proliferation and safety risks. Much of this material was to be found in a key nuclear weapon component known as a pit. The Department of Energy (DOE) planned to dispose of at least 34 metric tons of plutonium by fabricating it into mixed oxide (MOX) fuel for domestic nuclear reactors. To do so, DOE's National Nuclear Security Administration (NNSA) was constructing two facilities—a MOX Fuel Fabrication Facility (MFFF) and a Waste Solidification Building (WSB)—at the Savannah River Site in South Carolina. GAO was asked to assess the (1) cost and schedule status of the MFFF and WSB construction projects, (2) status of NNSA's plans for pit disassembly and conversion, (3) status of NNSA's plans to obtain customers for MOX fuel from the MFFF, and (4) actions that the Nuclear Regulatory Commission (NRC) and DOE had taken to provide independent nuclear safety oversight.

To develop its analysis, GAO reviewed NNSA documents and project data, toured DOE facilities, and interviewed officials from DOE, NRC, and nuclear utilities. The analysts found that the MFFF project was subjected to schedule delays stemming, in part, from the delivery of reinforcing bars that did not meet nuclear quality standards.

CASE STUDY 9: FROM *TRANSPORTATION WORKER IDENTIFICATION CREDENTIAL*, GAO-10-43, DECEMBER 10, 2009

The Transportation Worker Identification Credential (TWIC) program, managed by the Department of Homeland Security's (DHS) Transportation Security Administration

(TSA) and the U.S. Coast Guard, required maritime workers who accessed secure areas of transportation facilities to obtain a biometric identification card in order to gain access. A TWIC regulation set a national compliance deadline of April 15, 2009. In part to inform the development of a second TWIC regulation, TSA was conducting a pilot program to test the use of TWICs with biometric card readers. GAO was asked to evaluate (1) TSA's and the Coast Guard's progress and related challenges in implementing TWIC and (2) the management challenges, if any, that TSA, the Coast Guard, and DHS faced in executing the TWIC pilot test.

We reviewed TWIC enrollment and implementation documents and visited sites or interviewed officials at the seven pilot program sites. We found that TSA had made progress in incorporating management best practices to execute the TWIC pilot test, aimed at informing the Congress. But TSA faced two management challenges to ensure the successful execution of the test and the development of the second TWIC regulation. First, TSA faced problems in using the TWIC pilot schedule to both guide the pilot and accurately identify the pilot's completion date. Although TSA had improved its scheduling practices in executing the pilot, weaknesses remained, such as not capturing all pilot activities in the schedule. This could adversely affect the schedule's usefulness as both a management tool and a means of communication among pilot participants.

Second, shortfalls in planning for the TWIC pilot hindered TSA and the Coast Guard's efforts to ensure that the pilot (1) represented deployment conditions and (2) would yield the information needed—such as the operational effects of deploying biometric card readers and their costs—to accurately inform the Congress and develop the second regulation. This was partly because TSA and the Coast Guard had not developed an evaluation plan that fully identified the scope of the pilot or specified how information from the pilot would be analyzed. The current evaluation plan described data collection methods but did not identify the evaluation criteria and methodology for analyzing the pilot data once they were collected. A well-developed, sound evaluation plan would have helped TSA and the Coast Guard determine how the data were to be analyzed to measure the pilot's performance.

CASE STUDIES 11 AND 13: FROM *FAA ACQUISITIONS*, GAO-12-223, FEBRUARY 16, 2012

The Federal Aviation Administration (FAA), partnering with other federal agencies and the aviation industry, is implementing the Next Generation Air Transportation System (NextGen), a new satellite-based air traffic management system that will replace the current radar-based system and is expected to enhance the safety and capacity of the air transport system by 2025. In a review of 30 major ATC acquisition programs, all contributing to the transition to NextGen, GAO found that costs for 11 of the 30 programs had increased from their initial estimates by a total of $4.2 billion and that 15 programs had been delayed. The 11 acquisitions accounted for over 60 percent of FAA's total acquisition costs ($11 billion of $17.7 billion) for the 30 programs. The 15 acquisitions, 10 of which also had cost increases, ranged from 2 months to more than 14 years and averaged 48 months.

To determine the extent to which selected ATC programs adhered to best practices for determining acquisition costs and schedules, we conducted an in-depth review of 4 of the 30 acquisition programs: the Automatic Dependent Surveillance-Broadcast (ADS-B) system, the Collaborative Air Traffic Management Technologies (CATMT) system, the System Wide Information Management (SWIM) system, and the Wide Area Augmentation System (WAAS). In addition to conducting interviews, we collected documentation and analyzed and summarized the views and information we collected. We also performed a schedule risk analysis of the WAAS program to determine the likelihood of the project's finishing on schedule.

FAA was not consistently following the characteristics of high-quality cost estimates and scheduling best practices for the four programs GAO analyzed. Regarding scheduling practices, most programs did not substantially or fully meet the majority of the 9 best practices GAO previously identified, including developing a fully integrated master schedule of all program activities and performing a schedule risk analysis. For example, without a schedule risk analysis, FAA is unable to predict, with any degree of confidence, whether the estimated completion dates are realistic.

FAA is implementing new processes and organizational changes to better manage acquisitions. However, by not consistently following the characteristics of high-quality cost estimate and scheduling best practices, FAA cannot provide reasonable assurance to the Congress and other stakeholders that NextGen and other ATC programs will avoid additional cost increases or schedule delays.

CASE STUDY 12: FROM *IMMIGRATION BENEFITS*, GAO-12-66, NOVEMBER 22, 2011

Each year, DHS's U.S. Citizenship and Immigration Services (USCIS) handles millions of paper applications for immigration benefits. In 2005, USCIS embarked on a major multiyear program meant to transform its paper process to one that would incorporate electronic application filing, adjudication, and case management. In 2007, we reported that USCIS planned, in the early stages of the program, to partially or fully meet best practices. In 2008, USCIS contracted with a solutions architect to help develop the new program.

We reviewed DHS's acquisition management policies and guidance; analyzed transformation program planning and implementation documents, such as operational requirements; compared schedule and cost information with our guidance for best practices; and interviewed USCIS officials. We found that USCIS was continuing to manage the program without specific acquisition management controls, such as reliable schedules, that would have detailed work to be performed by both the government and its contractor over the expected life of the program. As a result, USCIS did not have reasonable assurance that it could meet its future milestones. In particular, although USCIS had established schedules for the first release of the transformation program, our analysis showed that these schedules were unreliable because they did not meet best practices for schedule estimating.

CASE STUDY 14: FROM *COAST GUARD*, GAO-11-743, JULY 28, 2011

The Deepwater Program—the largest acquisition program in the Coast Guard's history—began in 1996 to recapitalize the Coast Guard's operational fleet, including ships, aircraft, and other supporting capabilities. In 2007, the Coast Guard took over the lead systems integrator role from the Integrated Coast Guard Systems, establishing a $24.2 billion overall program baseline.

We reviewed key Coast Guard documents and applied criteria from the GAO Cost Guide. We found that the estimated total acquisition cost of the Deepwater Program, based on approved program baselines as of May 2011, could have been as much as approximately $29.3 billion, or about $5 billion more than the $24.2 billion baseline DHS approved in 2007. However, we also found that two factors precluded a solid understanding of the program's true cost and schedule: (1) the Coast Guard had not yet developed revised baselines for all assets, including the Offshore Patrol Cutter—the largest cost driver in the program—and (2) the Coast Guard's most recent capital investment plan indicated further cost and schedule changes not yet reflected in the asset baselines. We also found that the reliability of the cost estimates and schedules for selected assets were undermined because the Coast Guard did not follow key best practices for developing them.

CASE STUDY 17: FROM *AVIATION SECURITY*, GAO-11-740, JULY 12, 2011

Explosives in baggage represent a continuing threat to aviation security. To detect explosives, TSA used the Electronic Baggage Screening Program (EBSP) for checked baggage. To identify and resolve threats in checked baggage, the program includes the use of the explosives detection system (EDS) in conjunction with explosives trace detection (ETD) machines.

We analyzed EDS requirements, compared the EDS acquisition schedule against our best practices, and interviewed DHS officials. We found that TSA had faced challenges in procuring the first 260 EDSs to meet 2010 requirements. For example, the danger associated with some explosives challenged TSA and DHS in developing simulants and collecting data on the explosives' physical and chemical properties. Vendors and agencies needed these data to develop detection software and test EDSs before acquisition. In addition, TSA's decision to pursue EDS procurement during data collection complicated both efforts and resulted in a delay of over 7 months for the 2011 EDS procurement.

TSA could have helped avoid additional schedule delays by completing data collection for each phase of the 2010 requirements before pursuing EDS procurement that met those requirements. Although TSA had established a schedule for the 2011 EDS procurement, the schedule did not fully comply with our best practices, and TSA had not developed a plan to upgrade its EDS fleet to meet the 2010 requirements.

Case Study 18: From *Polar-orbiting Environmental Satellites*, GAO-13-676, September 11, 2013

The National Polar-orbiting Operational Environmental Satellite System (NPOESS) program was planned to be a state-of-the-art, environment-monitoring satellite system that would replace two existing polar-orbiting environmental satellite systems. Managed jointly by the Department of Commerce's National Oceanic and Atmospheric Administration (NOAA), the U.S. Air Force, and the National Aeronautics and Space Administration (NASA), the program was considered critical to the nation's ability to maintain the continuity of data required for weather forecasting and global climate monitoring through 2026.

However, in the 8 years after the development contract was awarded in 2002, the NPOESS cost estimate had more than doubled—to about $15 billion—launch dates had been delayed by over 5 years, significant functionality had been removed from the program, and the program's tri-agency management structure had proven to be ineffective. Importantly, delays in launching the satellites put the program's mission at risk. To address these challenges, a task force led by the White House's Office of Science and Technology Policy (OSTP) reviewed the management and governance of the NPOESS program. In February 2010, OSTP's Director announced a decision to disband the NPOESS acquisition and, instead, have NOAA and DOD undertake separate acquisitions, with NOAA responsible for satellites in the afternoon orbit and DOD responsible for satellites in the early morning orbit. After that decision, NOAA began developing plans for the Joint Polar Satellite System (JPSS). In October 2011, the JPSS program successfully launched the Suomi National Polar-orbiting Partnership (S-NPP) demonstration satellite, the first in a series of satellites to be launched as part of NOAA's JPSS program.

Given the interest of the Congress in the progress NOAA had made on the JPSS program, our objectives were to evaluate (1) NOAA's progress in meeting program objectives of sustaining the continuity of the polar-orbiting satellite system through the S-NPP and JPSS satellites, (2) the quality of the JPSS program schedule, and (3) NOAA's plans to address potential gaps in polar satellite data.

We found that the JPSS program office did not yet have a complete integrated master schedule and weaknesses existed in component schedules. Specifically, the program established an integrated master schedule in June 2013 and was reporting a 70 percent confidence level in the JPSS-1 launch date. However, about one-third of the program schedule was missing information needed to establish the sequence in which activities occur. In addition, selected component schedules supporting the JPSS-1 satellite had weaknesses, including schedule constraints that had not been justified.

CASE STUDY 19: FROM 2010 *CENSUS*, GAO-10-59, NOVEMBER 13, 2009

To carry out the decennial census, the U.S. Census Bureau conducts a sequence of thousands of activities and numerous operations. As requested by the Congress, we examined its use of (1) scheduling tools to maintain and monitor progress and (2) two control systems key to field data collection—one to manage the work flow for paper-based operations, including nonresponse follow-up, and the other to manage quality control for two major field operations. We applied schedule analysis tools; reviewed the Bureau's evaluations, planning documents, and other documents on work flow management; and interviewed Bureau officials.

We found that as the Bureau carries out the census, its master schedule provides a useful tool to gauge progress, identify and address potential problems, and promote accountability. We also found that the Bureau's use of its master schedule generally follows leading scheduling practices, which allow such high-level oversight. However, the errors we found in the Bureau's master schedule were hindering its ability to identify the effects of activity delays and to plan for the unexpected.

APPENDIX X

EXPERTS WHO HELPED DEVELOP THIS GUIDE

The two lists in this appendix name the experts in the scheduling community, along with their organizations, who helped us develop this guide. This first list names significant contributors to the Schedule Guide. They attended and participated in numerous expert meetings, provided text or graphics, and submitted comments.

LIST 1

Organization	Expert	
ABBA Consulting	Wayne	Abba
Acumen	Brad	Arterbury
AECOM & PMI Scheduling President	Pradip	Mehta
AG Midgley Ltd.	Alan	Midgley
Air Force	Jennifer	Bowles
	John	Cargill
	Greg	Hogan
	Fred	Meyer
	Harold	Parker
	Donna	Rosenbaum
ARCADIS U.S., Inc.	Chris	Carson
	Mike	Debiak
ARES Software	Eden	Gold
Bechtel Jacobs	Darryl	Walker
Belstar, Inc. (former AACE President)	Osmund	Belcher
Booz Allen Hamilton	Javed	Hasnat
	Seth	Huckabee
Chartered Institute of Building	Saleem	Sakram
	Earl	Glenwright
Covarus, LLC	Raymond	Covert
CPIC Solutions Corp.	William	Mathis
David Consulting Group	Michael	Harris
Defense Acquisition University	Robert	Pratt
Defense Contract Management Agency	Erik	Berg

Organization	Expert	
Delta Consulting Group	Joseph	Perron
Department of Energy	Brian	Kong
	Reuben	Sanchez
Department of Homeland Security	Christine	Ketcham
	Robert	Uzel
Department of Justice	Lionel	Cares
	Teresa	Palmer
Department of the Treasury	Kimberly	Smith
Dunelm PMC, LLC	Keith	Corner
The Earned Value Group	Glenn	Counts
Ernst and Young	Kimberly	Hunter
ESPM, Inc.	Joe	Halligan
Federal Acquisition Certification Academy	Ben	Sellers
Federal Aviation Administration	Fred	Sapp
forProject Technology, Inc	Harry	Sparrow
General Services Administration	Bill	Hunt
Hill International	Keith	Pickavance
Hornbacher Associates and University of Pennsylvania	Keith	Hornbacher
Hulett & Associates, LLC	David	Hulett
The International Center for Scheduling, Inc. (ICS-Global)	Murray	Woolf
Independent consultant	Anthony	Corridore
	Joyce	Glenn
	Stephen	Lee
	Lawrence	Mugg
Institute for Defense Analyses	Tom	Coonce
KM Systems Group	Joe	Houser
Knowledge Advantage, Inc.	Scott	Gring
L-3 Stratis	Eric	Christoph
Legis Consultancy, Inc.	Patrick	Ray
Lexmark International, Inc.	Don	Green
Longview-FedConsulting JV	Charles	Cobb
	Sandra	Marin
Ludwig Consulting Services, LLC	Joyce	Ludwig
Management Technologies	Ray W.	Stratton
MBP	Jill	Hubbard
	Niyi	Ladipo

Organization	Expert	
MCR, LLC	Neil	Albert
	Susan	Barton
	Brian	Evans
	Jay	Goldberg
Missile Defense Agency	David	Anderson
	Ken	Twining
MITRE	Clarke	Thomason
	Nathan	Welch
National Aeronautics and Space Administration	Jimmy	Black
	Kristen	Kehrer
	Jerald	Kerby
National Science Foundation	Patrick	Haggerty
Naval Air Systems Command	John	Scaparro
Naval Center for Cost Analysis	Duncan	Thomas
Navigant	John	Livengood
Northrop Grumman	Raymond	Bollas
	Anthony	Claridge
	Gay	Infanti
Olde Stone Consulting, LLC	John	Driessnack
Oracle	Christopher	Sala
	Kristy	Tan
Parsons Brinckerhoff	Marie	Gunnerson
PMFocus	Dan	Patterson
Price Waterhouse Coopers	Jennifer	Mun
ProChain Solutions, Inc.	Rob	Newbold
Professional Project Management Services	Mike	Stone
Project & Cost Management Consulting Services	Christopher	Gruber
Project Management Consultant	Shashi	Khanna
Project Time & Cost	Michael	Nosbisch
	Chris	Watson
PT Mitratata Citragraha	Paul	Giammalvo
R. J. Kohl & Associates	Ronald	Kohl
Raytheon	Joshua	Anderson
	Warren	Kline
	Joseph	Kusick
ServQ and the University of Bristol	Andrew	Crossley
SFB PM Consulting	Stephen	Bonk

Organization	Expert	
SRA	Shobha	Mahabir
Tecolote Research, Inc.	Mike	Dalessandro
	Darren	Elliott
	Anthony	Harvey
	Greg	Higdon
	James	Johnson
	Linda	Milam
	Alfred	Smith
Trauner Consulting Services, Inc.	Scott	Lowe
VARiQ	Christopher	Ditta
Walter Majerowicz Consulting	Walter	Majerowicz

Source: GAO | GAO-16-89G.

This second list names those who generously donated their time to reviewing this guide in its various stages and who provided feedback.

LIST 2

Organization	Expert	
Accenture Federal Services	Tom	Maraglino
	Nick	Mark
	Mark	Nickolas
	Patrick	South
	Phil	Wood
Air Force	Shannon	House
	Dolores	LaGuarde
	Dennis	Rackard
Army	Sean	Vessey
AzTech International	Luis	Contreras
	Timothy	Fritz
	James	Ivie
	Zachary	Lindemann
	Dave	Rutter
	Blaine	Schwartz
	Joy	Sichveland
Barrios Technology	Patrick	McGarrity

Organization	Expert	
Battelle	Bill	Altman
	Chuck	Chapin
BIA	Vivian	Deliz
Boeing	Daniel	Dassow
	Kirk	Kotthoff
	Barbara	Park
Booz Allen Hamilton	Calvin	Speight
Chevo Consulting	Cyndy	Iwan
	Michelle	Powell
Cobec Consulting, Inc.	Dan	French
D&G	Chris	Alberts
David Consulting Group	David	Herron
Davis Langdon	Peter	Morris
Defense Contract Management Agency	James	Baber
	Marvin	Charles
	Donna	Holden
	Alexander	Schostag
Department of Commerce	Alpha	Bailey
	Peggy	Fouts
	Maria	Sims
Department of Education	Tauqir	Jilani
Department of Energy	Fredericka	Baker
	Richard	Couture
	Ronald	Lile
	Robert	Loop
	Victoria	Premaza
	Autar	Rampertaap
	Karen	Urschel
Department of Health and Human Services	Kimberly	Crenshaw
	Rita	Warren
Department of Homeland Security	Boris	Blechman
	Jose	Christian
	Michael	DiVecchio
	Katie	Geier
	James	Mararac
	Steve	Nakazawa
	Lauren	Riner

Organization	Expert	
	William	Taylor
	Michael	Zaboski
Department of Justice	Anthony	Burley
	Bellorh	Byrom
	Tapan	Das
	Bryce	Mitchell
Department of Transportation	Traci	Stith
Federal Aviation Administration	Sharon	Boddie
	William	Russell
Fluor Corp.	Eric	Marcantoni
Galorath, Inc.	Daniel	Galorath
	Bob	Hunt
General Services Administration	Patrick	Plunket
	Gene	Ransom
	John	Ray
George Washington University	Homayoun	Khamooshi
Humphreys and Associates	Yancy	Qualls
HII-Ingalls Shipbuilding, Inc.	Dan	Burke
Idaho National Laboratory	Rick	Staten
Independent Consultant	Sheila	DeBardi
	Gerard	Jones
	John	Pakiz
Internal Revenue Service	Donald	Moushegian
iSystems Group, Inc.	Andrew	Lovorn
ITG	Judy	Rexin
ITT Exelis	Brenda	Malmberg
	Bill	Mendolsohn
Kaiser Permanente	Kathy	Cook
KPMG	Mark	Hogenmiller
Legis Consultancy, Inc.	Dave	Smart
Lockheed Martin	Marcy	Barlett
	James	Fieber
Management & Aviation	Sven	Antvik
MCR, LLC	David	Treacy
MDS	Mark	Ives
MITRE	Raj	Agrawal
	Charlie	Dobbs

Organization	Expert	
	Eric	Fisher
	Daniel	Harper
	Richard	Riether
	Victoria	Smith
Monte Ingram & Associates, LLC	Monte	Ingram
National Aeronautics and Space Administration	Heidemarie	Borchardt
	Barbara	Carroon
	Claude	Freaner
	Carol	Grunsfeld
	David	Hall
	Arnold	Hill
	Jerry	Holsomback
	Zach	Hunt
	James	Johnson
	Fred	Kuo
	Patrick	Maggarity
	Nakia	Marks
	Arlene	Moore
	Ken	Poole
	Johnnetta	Punch
	Yvonne	Simonsen
	Mike	Soots
	Steve	Sterk
	Steve	Wilson
National Defense Industrial Association	Chris	Hassler
	Joan	Ugljesa
Naval Air Systems Command	Bruce	Koontz
	Jeffrey	Upton
Naval Center for Cost Analysis	Tim	Lawless
Naval Sea Systems Command	Hershel	Young
Northrop Grumman	Jim	Chappell
	Josh	Dornan
	Stanton	Smith
Oak Ridge National Laboratory	Bill	Toth
Office of Management and Budget	Jim	Wade
Office of the Secretary of Defense	Joseph	Beauregard
	David	Nelson

Organization	Expert	
Peace Corps	Robyn	Wiley
Pratt & Whitney Rocketdyne	Gregory	Manley
	Fernando	Vivero
PRICE Systems, LLC	Zach	Jasnoff
Project & Program Management Solutions	Matthew	Morris
Professional Project Services (Pro2Serve)	Greg	Dowd
QS Requin Corp.	Alexia	Nalewaik
Raytheon	Jeff	Poulson
The Rehancement Group, Inc.	Peter	Chrzanowski
Ron Winter Consulting	Ron	Winter
SAIC	Ralph	Justus
	Laura	Mraz
SETA Contractor	John	Roland
SM&A	Mark	Infanti
Systems Made Simple	Ron	Lattomus
Technomics	Peter	Braxton
	Brian	Octeau
Tecolote Research, Inc.	Michael	Harrison
	Bill	Rote
United States Capitol Police	Ken	Sragg

Source: GAO | GAO-16-89G.

APPENDIX XI

GAO Contacts and Staff Acknowledgments

GAO Contact

Timothy M. Persons, Ph.D., Chief Scientist, at (202) 512-6412 or personst@gao.gov

Other Leadership on This Project

Jason T. Lee, Assistant Director, Applied Research and Methods (ARM), and Karen Richey, Assistant Director, also in ARM

Key Contributors

Pille Anvelt, Visual Communications Analyst

Ellen Arnold-Losey, Visual Communications Analyst

Mathew Bader, Information Technology Analyst

Brian Bothwell, Senior Cost Analyst

Amy Bowser, Senior Attorney

Carol R. Cha, Director, Information Technology Issues

Juaná Collymore, Senior Cost Analyst

Kimani Darasaw, Analyst

Jennifer Echard, Senior Cost Analyst

Eric Hauswirth, Visual Communications Analyst

Kaelin Patrick Kuhn, Senior Analyst

Josh Leiling, Information Technology Analyst

Jennifer Leotta, Senior Cost Analyst

Les Locke, Physical Infrastructure Analyst

Constantine (Dino) Papanastasiou, Information Technology Analyst

Penny Pickett, Ph.D., Senior Communications Analyst

GLOSSARY

Backward pass	A calculation in a schedule network that determines late start dates by subtracting durations from late finish dates
Baseline schedule	Represents the original configuration of the program plan and signifies the consensus of all stakeholders regarding the required sequence of events, resource assignments, and acceptable dates for key deliverables
Basis document	A single document that defines the organization of the IMS, describes the logic of the network, describes the basic approach to managing resources, and provides a basis for all parameters used to calculate dates
Consolidated schedule	An IMS that aggregates multiple project files in a single master file for reporting or management purposes, even if those projects are immaterially related. Also known as a portfolio schedule, although portfolio schedule and consolidated schedule are often synonymous with IMS
Contingency	A margin or a reserve of extra time to account for known and quantified risks and uncertainty
Critical activity	An activity on the critical path. When the network is free of date constraints, critical activities have zero float, and therefore any delay in the critical activity causes the same day-for-day amount of delay in the program forecast finish date
Critical path	The longest continuous sequence of activities in a schedule. Defines the program's earliest completion date or minimum duration
Dangling logic	Scheduling logic that is not properly tied to an activity's start or end date. Also referred to as hanging logic
Date constraint	An override of the calculated start or finish dates of activities by imposing calendar restrictions on when an activity can begin or end
Detail activity	Activities at the lowest level of the WBS representing the performance of actual discrete work that is planned in the project. Logically related paths of detail activities are linked to milestones to show the progression of work that is planned
Detail schedule	The lowest level of schedule. The detail schedule lays out the logically sequenced day-to-day effort to reach program milestones
Deterministic critical path	The critical path as defined by the initial or current set of inputs in the schedule model
Duration	The estimated time required to complete an activity—the time between its start and finish. Durations are expressed in business units, such as working days, and are subjected to the project calendar
Finish no earlier than (FNET)	A date constraint that schedules an activity to finish on or after a certain date. That is, FNET constraints prevent an activity from finishing before a certain date. Also called finish on or after constraints
Finish no later than (FNLT)	A date constraint that schedules an activity to finish on or before a certain date. That is, FNLT constraints prevent an activity from finishing after a certain date. FNLT constraints are also called finish on or before constraints
Finish-to-finish (F–F)	A logic relationship that dictates that a successor activity cannot finish until the predecessor activity finishes

Finish-to-start (F–S)	A logic relationship that dictates that a successor activity cannot start until the predecessor activity finishes
Float	See total float or free float. Also referred to as slack
Forward pass	A calculation in a schedule network that determines the early start and early finish times for each activity by adding durations successively through the network, starting at day one. The forward pass will derive the total time required for the entire project by calculating the longest continuous path through the network
Fragnet	A fragmentary, or subordinate, network that represents a sequence of activities typically related to repetitive effort. Subordinate networks can be inserted into larger networks as a related group of activities
Free float	The portion of an activity's total float that is available before the activity's delay affects its immediate successor. Depending on the sequence of events in the network, an activity with total float may or may not have free float
Giver/receiver	Represents dependencies between schedules, such as hand-offs between integrated product teams and delivery and acceptance of government-furnished equipment
Horizontal traceability	Demonstrates that the overall schedule is rational, has been planned in a logical sequence, accounts for the interdependence of detailed activities and planning packages, and provides a way to evaluate current status. Schedules that are horizontally traceable depict logical relationships between different program elements and product handoffs
Integrated master schedule (IMS)	A program schedule that includes the entire required scope of effort, including the effort necessary from all government, contractor, and other key parties for a program's successful execution from start to finish. The IMS should consist of logically related activities whose forecasted dates are automatically recalculated when activities change. The IMS includes summary, intermediate, and detail-level schedules
Intermediate Schedule	The intermediate schedule includes all information displayed in the summary schedule, as well as key program activities and milestones that show the important steps in achieving high-level milestones
Lag	Denotes the passage of time between two activities. Lags simply delay the successor activity—no effort or resources are associated with this passage of time
Lead	A negative lag used to accelerate a successor activity. Leads imply the unusual measurement of negative time and exact foresight about future events
Level-of-effort (LOE) activity	An activity that represents effort that has no measurable output and cannot be associated with a physical product or defined deliverable. LOE activities are typically related to management and other oversight that continues until the detailed activities they support have been completed
Longest path	Theoretically, the longest path is equal to the critical path. As a schedule becomes more complex, total float values may not necessarily represent a true picture of schedule flexibility. In those cases, the longest path is the sequence of activities directly affecting the estimated finish date of the key milestone, ignoring the presence of any date constraints
Merge bias	The additional risk at points in the schedule where parallel paths merge
Milestone	Points in time that have no duration but that denote the achievement or realization of key events and accomplishments such as program events or contract start dates. Because milestones lack duration, they do not consume resources
Must finish on (MFON)	A date constraint that schedules an activity to finish on a certain date. That is, MFON constraints prevent an activity from finishing any earlier or later than a certain date, thereby overriding network logic. MFON constraints are also called mandatory finish constraints
Must start on (MSON)	A date constraint that schedules an activity to start on a certain date. That is, MSON constraints prevent an activity from starting any earlier or later than a certain date, thereby overriding network logic. MSON constraints are also called mandatory start constraints
Near-critical activity	An activity with total float within a narrow range of the critical path. Near-critical activities can quickly become critical if their small amount of total float is used up in a delay
Out-of-sequence logic	The result of progress on an activity performed in a different order from that originally planned

Path convergence	Several parallel activities joining with a single successor activity
Performance measurement baseline	A time-phased budget plan for accomplishing work. Performance is measured against the PMB
Predecessor	Activities that are logically related within a schedule network are referred to as predecessors and successors. A predecessor activity must start or finish before its successor
Probabilistic branching	The addition of new activities in a schedule that occur only with some probability. Probabilistic branching is used to model the random choice between two alternatives
Progress override	When out-of-sequence progress occurs, managers and schedulers may choose to override the existing network logic. Work on the activity that began out of sequence is permitted to continue, regardless of original predecessor logic. Actual progress in the field supersedes the plan logic, and work on the out-of-sequence activity continues
Resource	Anything required to perform work, such as labor, materials, travel, and facilities
Resource leveling	Adjusts the scheduled start of activities or the work assignments of resources to account for their availability. Leveling is used primarily by the organization that has control of the resources to smooth spikes and troughs in resource demands created by the sequencing of activities in the schedule network
Retained logic	When out-of-sequence progress occurs, managers and schedulers may choose to retain existing network logic. Work on the activity that began out of sequence is stopped until its predecessor is completed. As much as possible of the original network logic is preserved because the remainder of the out-of-sequence activity is delayed until the predecessor finishes, to observe its original sequence logic
Risk	An uncertain event that could affect the program positively or negatively. Risk and its outcomes can be quantified in some definite way
Rolling wave planning	The incremental conversion of work from planning packages to detailed work packages. Rolling wave planning with portions of effort that align to significant program increments, blocks, or updates is sometimes referred to as block planning
Schedule narrative	A document that accompanies the updated schedule to provide a log of changes and their effect, if any, on the schedule time
Schedule risk analysis	An analysis that uses statistical techniques to predict a level of confidence in meeting a program's completion date. A schedule risk analysis focuses on uncertainty and key risks and how they affect the schedule's activity durations
Slack	A synonym for float
Start no earlier than (SNET)	A date constraint that schedules an activity to start on or after a certain date, even if its predecessors start or finish earlier. That is, SNET constraints prevent an activity from beginning before a certain date. SNET constraints are also called start on or after constraints
Start no later than (SNLT)	A date constraint that schedules an activity to start on or before a certain date. That is, SNLT constraints prevent an activity from starting any later than a certain date. SNLT constraints are also called start on or before constraints
Start-to-finish (S–F)	A theoretical logic relationship that has the bizarre effect of directing a successor activity not to finish until its predecessor activity starts
Start-to-start (S–S)	A logic relationship that dictates that a successor activity cannot start until the predecessor activity starts
Statement of work (SOW)	Defines, either directly or by reference to other documents, performance requirements for a contractor's effort. The SOW specifies the work to be done in developing the goods or services to be provided by a contractor
Status date	Denotes the date of the latest update to the schedule and thus defines the demarcation between actual work performed and remaining work. Also called a data date or time-now date
Statusing	The process of updating a plan with actual dates, logic, and progress and adjusting forecasts of the remaining effort

Successor	Activities that are logically related within a schedule network are referred to as predecessors and successors. A predecessor activity must start or finish before its successor
Summary activity	A grouping element that shows the time that activities of lower levels of detail require. Summary activities derive their start and end dates from lower-level activities
Summary schedule	Provides a strategic view of the activities and milestones necessary to start and complete a program. Summary schedules are roll-ups of lower-level intermediate and detail schedules
Total float	The amount of time an activity can be delayed or extended before delay affects the program's finish date. If positive, it indicates the amount of time that an activity can be delayed without delaying the program's finish date. If negative, it indicates the amount of time that must be recovered so as not to delay the program's finish date beyond the constrained date. Zero total float means that any amount of activity delay will delay the program finish date by an equal amount
Vertical traceability	Demonstrates the consistency of dates, status, and scope requirements between different levels of a schedule—summary, intermediate, and detailed. When schedules are vertically traceable, lower-level schedules are clearly consistent with upper-level schedule milestones, allowing for total schedule integrity and enabling different teams to work to the same schedule expectations
Work breakdown structure	Deconstructs a program's end product into successively greater levels of detail until the work is subdivided to a level suitable for management control
Work package	An activity or grouping of activities at the lowest level of the work breakdown structure, where work is planned and progress is measured

REFERENCES

AFMC (Air Force Materiel Command). *Integrated Master Plan and Schedule Guide.* AFMC Pamphlet 63-5. Wright-Patterson AFB, Ohio: 2005.

Ambriz, Rodolfo. *Dynamic Scheduling with Microsoft® Project 2007.* Fort Lauderdale, Fla.: International Institute for Learning, 2008.

Ambriz, Rodolfo, and John White. *Dynamic Scheduling with Microsoft® Project 2010.* Fort Lauderdale, Fla.: International Institute for Learning, 2011.

Anderson, Joshua, and Jeff Upton. "Unleashing the Predictive Power of the Integrated Master Schedule." *Defense AT&L*, January-February 2012: 34–38.

Anderson, Mark I. *Recovery Schedules.* Rockville, Md.: Warner Construction Consultants, 2006.

Antvik, Sven, and Håkan Sjöholm. *Project Management and Methods (Projekt–ledning och metoder).* Västerås, Sweden: Projektkonsult Håkan Sjöholm AB, 2007.

Bachman, David. *Better Schedule Performance Assessments Derived from Integrated Master Plan-Referenced Schedule Metrics.* Defense Acquisition University, October 2011.

Bolinger, Paul. *Schedule Analysis of the Integrated Master Schedule.* Orange, Calif.: Humphreys & Associates, May 2008.

Book, Stephen A. "Schedule Risk Analysis: Why It Is Important and How to Do It." Presentation to the International Society of Parametric Analysts and the Society of Cost Estimating and Analysis. MCR. Denver, Colorado: June 14–17, 2005.

Chartered Institute of Building. *Guide to Good Practice in the Management of Time in Complex Projects.* Oxford: Wiley-Blackwell, 2011.

Cooper, Sue L. "Basic Schedule Analysis." Presentation to the International Society of Parametric Analysts and the Society of Cost Estimating and Analysis. Boeing, Denver, Colorado:, June14–17, 2005.

Defense Acquisition University (DAU). *ACC Practice Center—Risk Management.* Fort Belvoir, Va.: December 2011.

DAU. *Defense Acquisition Guidebook Chapter 14—Acquisition of Services*, accessed May 20, 2015, http://acc.dau.mil/communitybrowser.aspx?id=490645.

Defense Contract Management Agency (DCMA). *Earned Value Management System Compliance Reviews*, DCMA-INST 208. Washington, D.C.: April 9, 2014.

DCMA. *Finding the Program Critical Path*, EVC-100 Rev1. Fort Lee, Va: November 20, 2012.

DCMA. *Integrated Master Plan/Integrated Master Schedule Basic Analysis*. Washington, D.C.: Nov. 21, 2009.

DCMA. *Open Plan Integrated Master Schedule Assessment Guide*, EVC-102 Rev8. Chester, Va: March 27, 2013.

DCMA. *Overview: 14 Point Assessment*, EVC-104 Rev1. Fort Lee, Va: n.d.

DCMA. *Primavera® Integrated Master Schedule Assessment Guide*, EVC-103 Rev7. Chester, Va: March 27, 2013.

DCMA. *Project® Integrated Master Schedule Assessment Guide*, EVC-101 Rev11. Chester, Va: March 27, 2013.

DCMA. *Schedule Margin*, EVC-106 Rev2. Alexandria, Va: March 14, 2013.

Defense Systems Management College. *Schedule Guide for Program Managers*. Fort Belvoir, Va.: October 2001.

Department of Defense (DOD). *Department of Defense Standard Practice: Work Breakdown Structures for Defense Materiel Items,* MIL-STD-881C. Washington, D.C.: October 3, 2011.

DOD. *Integrated Master Plan and Integrated Master Schedule*. Washington, D.C.: October 21, 2005.

DOD. *Integrated Master Schedule Data Item Description,* DI-MGMT-81650. Washington, D.C.: March 30, 2005.

DOD. *Integrated Program Management Report*, DI-MGMT-81861. Washington, D.C.: June 20, 2012.

DOD. *Integrated Program Management Report Implementation Guide*, Washington, D.C.: January 24, 2013.

DOD. *Joint Agency Cost Schedule Risk and Uncertainty Handbook*. Washington, D.C.: March 12, 2014.

Department of Homeland Security. *Scheduling Handbook*. Washington, D.C.: August 29, 2014.

Douglas, Edward E. III. *Documenting the Schedule Basis,* AACE International Recommended Practice 38R-06. Morgantown, W. Va.: 2009.

Federal Aviation Administration. *Acquisition Management Policy.* Washington, D.C.: January 2012.

Government Accountability Office (GAO). *Air Traffic Control Modernization: Management Challenges Associated with Program Costs and Schedules Could Hinder NextGen Implementation,* GAO-12-223. Washington, D.C.: February 16, 2012.

GAO. *Arizona Border Surveillance Technology Plan: Additional Actions Needed to Strengthen Management and Assess Effectiveness,* GAO-14-368. Washington, D.C.: March 3, 2014.

GAO. *Aviation Security: TSA Has Enhanced Its Explosives Detection Requirements for Checked Baggage, but Additional Screening Actions Are Needed,* GAO-11-740. Washington, D.C.: July 11, 2011.

GAO. *2010 Census: Census Bureau Has Made Progress on Schedule and Operational Control Tools, but Needs to Prioritize Remaining System Requirements,* GAO-10-59. Washington, D.C.: November 13, 2009.

GAO. *Coast Guard: Action Needed as Approved Deepwater Program Remains Unachievable,* GAO-11-743. Washington, D.C.: July 28, 2011.

GAO. *Cost Estimating and Assessment Guide,* GAO-09-3SP. Washington, D.C.: March 2009.

GAO. *DOD Business Systems Modernization: Air Force Business System Schedule and Cost Estimates,* GAO-14-152. Washington, D.C.: February 7, 2014.

GAO. *DOD Business Transformation: Improved Management Oversight of Business System Modernization Efforts Needed,* GAO-11-53. Washington, D.C.: October 7, 2010.

GAO. *Nuclear Nonproliferation: DOE Needs to Address Uncertainties with and Strengthen Independent Safety Oversight of Its Plutonium Disposition Program,* GAO-10-378. Washington, D.C.: March 26, 2010.

GAO. *Immigration Benefits: Consistent Adherence to DHS's Acquisition Policy Could Help Improve Transformation Program Outcomes,* GAO-12-66. Washington, D.C.: November 22, 2011.

GAO. *Nuclear Waste: Actions Needed to Address Persistent Concerns with Efforts to Close Underground Radioactive Waste Tanks at DOE's Savannah River Site,* GAO-10-816. Washington, D.C.: September 14, 2010.

GAO. *Polar Weather Satellites: NOAA Identified Ways to Mitigate Data Gaps, but Contingency Plans and Schedules Require Further Attention,* GAO-13-676. Washington, D.C.: September 11, 2013.

GAO. *Transportation Worker Identification Credential: Progress Made in Enrolling Workers and Activating Credentials but Evaluation Plan Needed to Help Inform the Implementation of Card Readers,* GAO-10-43. Washington, D.C.: November 18, 2009.

GAO. *VA Construction: VA Is Working to Improve Initial Project Cost Estimates, but Should Analyze Cost and Schedule Risks,* GAO-10-189. Washington, D.C.: December 14, 2009.

Hillson, David. *Use a Risk Breakdown Structure (RBS) to Understand Your Risks,* Proceedings of the Project Management Institute Annual Seminars & Symposium. San Antonio, Texas: October 3–10, 2002.

Hulett, David T. *Advanced Project Scheduling and Risk Analysis.* Los Angeles, Calif.: Hulett and Associates, 2009.

Hulett, David T. *Advanced Quantitative Schedule Risk Analysis.* Los Angeles, Calif.: Hulett and Associates, 2007.

Hulett, David T. *Practical Schedule Risk Analysis.* Farnham, Surrey, U.K.: Gower, 2009.

Humphreys and Associates, Inc. *Project Management Using Earned Value.* Orange, Calif.: 2002.

International Institute for Learning, Inc. *Project Orange Belt® 2007.* New York: 2007.

National Aeronautics and Space Administration (NASA). *Cost Estimating Handbook.* Washington, D.C.: February 2015.

NASA. *Scheduling Management Handbook,* SP-2010-3403. Washington, D.C.: March 2011.

National Defense Industrial Association. *Planning and Scheduling Excellence Guide,* v2.0. Arlington, Va.: June 22, 2012.

Naval Air Systems Command (NAVAIR). *EVM Analysis Toolkit.* Washington, D.C.: April 14, 2009.

NAVAIR. *Integrated Master Schedule Guidebook.* Washington, D.C.: February 2010.

NAVAIR. *Integrated Project Management Toolkit*. Washington, D.C.: n.d.

NAVAIR. *Schedule Metrics Guide*. Washington, D.C.: August 8, 2013.

NAVAIR. *Schedule Risk Assessment Toolkit*. Washington, D.C.: September 27, 2010.

O'Brien, James J., and Fredric L. Plotnick. *CPM in Construction Management,* 6th ed. New York: McGraw-Hill, 2005.

Project Management Institute, Inc. *Practice Standard for Earned Value Management,* 2nd ed. Newtown Square, Pa.: 2011.

Project Management Institute, Inc. *Practice Standard for Project Risk Management*. Newtown Square, Pa.: 2009.

Project Management Institute, Inc. *Practice Standard for Scheduling,* 2nd ed. Newtown Square, Pa.: 2011.

U.S. Air Force. *Integrated Master Schedule Assessment Process*. Washington, D.C.: June 7, 2012.

U.S. Army Corps of Engineers. *Cost and Schedule Risk Analysis Guidance*. Draft. Washington, D.C.: May 17, 2009.

U.S. Immigration and Customs Enforcement. *Schedule Processes and Procedures,* v. 1.0. Washington, D.C.: March 28, 2012.

Uyttewaal, Eric. *Dynamic Scheduling with Microsoft® Project 2003*. Boca Raton, Fla.: International Institute for Learning, 2005.

VA (Department of Veterans Affairs). *Master Construction Specifications—Architectural and Engineering CPM Schedules,* 01 32 16.01. Washington, D.C.: n.d.

VA. *Master Construction Specifications—Network Analysis Schedules,* 01 32 16.13. Washington, D.C.: n.d.

VA. *Master Construction Specifications—Project Schedules (Design/Build),* 01 32 16.17. Washington, D.C.: n.d.

VA. *Master Construction Specifications—Project Schedules (Design-Build Only),* 01 32 16.16. Washington, D.C.: n.d.

VA. *Master Construction Specifications—Project Schedules (Small Projects – Design/Bid/Build),* 01 32 16.15. Washington, D.C.: n.d.

Weaver, Patrick. *Calculating and Using Float,* November, 2009. Accessed August 25, 2015, http://www.mosaicprojects.com.au/Resources_Papers_110.html.

(460623)

www.ingramcontent.com/pod-product-compliance
Lightning Source LLC
Chambersburg PA
CBHW081719220526
45468CB00008B/1902